Energy, Climate and the Environment

Series Editor

David Elliott
The Open University
Milton Keynes
United Kingdom

Aim of the Series

The aim of this series is to provide texts which lay out the technical, environmental and political issues relating to proposed policies for responding to climate change. The focus is not primarily on the science of climate change, or on the technological detail, although there will be accounts of this, to aid assessment of the viability of various options. However, the main focus is the policy conflicts over which strategy to pursue. The series adopts a critical approach and attempts to identify flaws in emerging policies, propositions and assertions. In particular, it seeks to illuminate counter-intuitive assessments, conclusions and new perspectives. The intention is not simply to map the debates, but to explore their structure, their underlying assumptions and their limitations. The books in this series are incisive and authoritative sources of critical analysis and commentary, clearly indicating the divergent views that have emerged whilst also identifying the shortcomings of such views. The series does not simply provide an overview, but also offers policy prescriptions.

More information about this series at
http://www.springer.com/series/14966

Richard H. Rosenzweig

Global Climate Change Policy and Carbon Markets

Transition to a New Era

Richard H. Rosenzweig
Washington, District of Columbia
US

ISBN 978-1-349-71959-4   ISBN 978-1-137-56051-3   (eBook)
DOI 10.1057/978-1-137-56051-3

Library of Congress Control Number: 2016947273

Cover image © Frans Lanting Studio / Alamy Stock Photo

Printed on acid-free paper

This Palgrave Macmillan imprint is published by Springer Nature
The registered company is Macmillan Publishers Ltd. London

Author's Preface

I decided to write this book approximately 25 years after getting involved in the climate change issue. In the early days, my interest in the issue was based on its multiple dimensions. The many characteristics that distinguished climate change from any other environmental and energy issue also increased the challenge of solving it. Impacts of such a change would affect all countries in the world, the rich and the poor alike, and the greenhouse gas (GHG) emissions that contributed to climate change were intrinsic to nearly every aspect of the economic activity. And sources of energy which powered the global economy were created predominantly by fossil fuels that emitted large quantities of carbon dioxide (CO_2), the dominant GHG.

I was also intrigued by the temporal dimensions of climate change and the characteristics of GHG emissions. The climate system was affected by cumulative GHG emissions over decades, measured in concentrations that remained in the atmosphere for long periods of time and not annual emissions. Achieving a specified concentration ceiling to reduce the risks and impacts of climate change required limiting GHG emissions to a fixed amount during the century. Strategies and policies needed to be put in to place to reduce GHG emissions throughout the twenty-first century, with a goal toward decarbonizing the energy system.

Because of these dynamics, domestic and international policy-makers would not be able to solve climate change by passing one piece of legislation or agreeing to a treaty. Instead, the entire economic system and the vast fossil fuel energy infrastructure that drove it, valued in terms of trillions of dollars, would need to be remade into a clean modern network that emitted little or no GHGs, and particularly CO_2. This would be an enormous, complicated, and expensive undertaking.

The US was at the center of the issue. It was the largest economy and emitter of GHGs in the world. The power system, its largest emitting sector, was dominated by coal, the most carbon-intensive fuel, and one with hundreds of years of reserves in the ground. And the transportation sector was almost entirely dependent on oil, the second most carbon-intensive fuel. So, it seemed apparent to me that the US political system would organize itself to reduce GHG emissions at home and play a lead role in the international community's efforts to develop a global response. Policy-makers started to put building blocks in place to do so. One of the last, if not the last, amendment debated on the floor of the House of Representatives on the Clean Air Act Amendments of 1990 was the requirement for electric utilities to report their CO_2 emissions. It was adopted, but not before it created a firestorm of opposition. Following that, a provision was included in the Energy Policy Act of 1992 that established a system allowing companies to report actions taken to reduce their GHG emissions. At the global level, the world also adopted the United Nations Framework Convention on Climate Change (UNFCCC), and while modest, it appeared to be a framework the international community could build on.

After coming into office, the Clinton Administration, in which I served as the Chief of Staff of the Department of Energy (DOE), quickly proposed an energy tax and committed to reduce US GHG emissions. It also played a lead role in the negotiations that culminated in the Kyoto Protocol (KP), the first agreement committing industrial nations to limit their GHG emissions and which created a market designed to stimulate investment in activities to do so. None of this ended well in the US. The tax proposed by President Clinton was soundly defeated and the US Senate never voted on ratification of the KP, its seeds of defeat having been sown years earlier. And the debate in the US over climate change during this time was dominated by disagreement as to whether a problem even existed, not over solutions to address its causes.

The Clinton Administration was replaced by the Administration of George W. Bush. It quickly stated its opposition to the KP and put the world on notice that the US had little to no intent of reducing its GHG emissions. This rallied the world to take the actions necessary for the KP to take effect and to use the markets as the primary mechanism to reduce global GHG emissions. In 2000, recognizing that there would be no domestic climate change response for many years, I departed the policy world to gain experience with the markets. I ended up as the Chief Operating Officer (COO) of Natsource, a company which became the largest buyer of carbon credits in the world through 2007. By the election of 2008, with GHG emissions growing in the US as well as globally, it was known that the KP was unworkable, because it covered only a sliver of global emissions, and that a new approach was needed. The markets created by the KP and other policies were in tatters; their performance was adversely affected by market design, market administration, and the severe economic downturn.

It was at this point that Barack Obama became the president of the US. He had committed to reduce GHG emissions at home and to restore US leadership in the international climate change negotiations. Unfortunately, his support of legislation to create an economy-wide national cap-and-trade system to control GHG emissions and to remake the nation's energy system failed in 2010, requiring an entirely new approach at home. During the same period, the international community was engaged in efforts to develop a successor to the KP. As I write this, the US has established goals for reducing its GHG emissions and finally put policies in place in an attempt to achieve them. And nearly 200 countries around the world agreed to a successor treaty to the KP in December of 2015 to reduce GHG emissions which had grown by 40 % from its 1990 base year to the end of its emissions reduction period in 2012. Both these efforts are in their nascent stages, and it is too early to determine whether they will succeed. The bottom line is that significant reductions in global GHG emissions, by 40 to 70 % by 2050 and to virtually net zero by 2100, are necessary to achieve long-term climate policy objectives.

About 25 years ago, I would have bet that by now the US and the world would have fashioned the necessary policies to address climate

change, particularly as so much was learned about its causes and consequences. I wrote this book with one goal in mind: to use my government
and business experience to contribute to the ongoing efforts to create
an enduring, effective response to global climate change. I attempt to
do this by describing the policies proposed and adopted during the last
25 years, and assessing their performance. I use the lessons drawn from
this exercise to recommend criteria to guide future policy-making efforts
and policies that can slow GHG emissions. I hope you enjoy the book
and I do welcome any feedback you may have.

Washington, DC, US   Richard H. Rosenzweig

Series Editor's Preface

Concerns about the potential environmental, social, and economic impacts of climate change have led to a major international debate over what could and should be done to reduce emissions of greenhouse gases (GHGs). There is still a scientific debate over the likely scale of the severity of climate change, and the complex interactions between human activities and climate systems, but, global average temperatures have risen and the cause is almost certainly the observed build up of atmospheric GHGs.

Whatever we now do, there will have to be a lot of social and economic adaptation to climate change—preparing for increased flooding and other climate-related problems. However, the more fundamental response is to try to reduce or avoid the human activities that are causing climate change. That means, primarily, trying to reduce or eliminate emission of GHGs from the combustion of fossil fuels. Given that around 80% of the energy used in the world at present comes from these sources, this will be a major technological, economic, and political undertaking. It will involve reducing demand for energy (via lifestyle choice changes—and policies enabling such choices to be made), producing and using whatever energy we still need more efficiently (getting more from less), and supplying the reduced amount of energy from non-fossil sources (basically switching over to renewables and/or nuclear power).

Each of these options opens up a range of social, economic, and environmental issues. Industrial society and modern consumer cultures have been based on the ever-expanding use of fossil fuels, so the changes required will inevitably be challenging. Perhaps equally inevitable are disagreements and conflicts over the merits and demerits of the various options and in relation to strategies and policies for pursuing them. These conflicts and associated debates sometimes concern technical issues, but there are usually also underlying political and ideological commitments and agendas which shape, or at least color, the ostensibly technical debates. In particular, at times, technical assertions can be used to buttress specific policy frameworks in ways which subsequently prove to be flawed.

The aim of this series is to provide texts which lay out the technical, environmental, and political issues relating to the various proposed policies for responding to climate change. The focus is not primarily on the science of climate change, or on the technological detail, although there will be accounts of the state of the art, to aid assessment of the viability of the various options. However, the main focus is the policy conflicts over which strategy to pursue. The series adopts a critical approach and attempts to identify flaws in emerging policies, propositions, and assertions.

The present text certainly looks at an area where there is no shortage of disagreements about policies—the attempt to develop carbon trading systems and carbon markets as a response to climate change. The author was involved with US policy formation and practice in this area and brings an insider's view to the debate on how to proceed in future. Carbon trading is seen by some as a market mechanism which ought to appeal to those on the political right, but it is also inevitably seen as a device for reducing fossil fuel use, and thus as suspect for those who do not believe that climate change is man-made. The polarization of views seem very strong in the US, the main focus of this book, less so in the EU, but, overall, real or contrived uncertainties about climate issues are making it hard to adopt the radical positions that some feel are needed to limit climate impacts. The approaches that have been adopted so far have clearly not been very successful: despite the KP and the attempt to use carbon markets to stimulate change, emissions have in general continued to rise. Given this situation, this book argues that it may be wise, or at least necessary, to adopt less ambitious approaches and more modest, targeted policies. That, it claims, may be more successful, and in terms of fighting climate change, policy successes are urgently needed.

David Elliott

Acknowledgments

This book has been both a labor of love and a significant challenge for me. I had wanted to write the book two to three years earlier, but the sudden death of a family member caused a delay in the project.

Like all challenges, I could not have succeeded without the encouragement and support of family members, friends, and colleagues. I want to acknowledge those that follow.

First, my family. I want to thank my mom who gave me my work ethic and was always there for me. I could not have completed the project without the support of my wife and partner, Sarah Wade. She encouraged, pushed, and cajoled me to take on this project and provided me with the confidence and support necessary to complete the book. Sarah also served as the primary editor—which is not an easy job. I also want to thank my son, Joel, for becoming the person he has. And last but not the least, thanks to my family of friends who put up with my complaints and encouraged me throughout.

I want to acknowledge my colleagues who contributed to the formulation and writing of the book. First, thanks to Dr. James A. Thurber of American University, whom I've known since 1983 when I walked into my first graduate school class. Jim is a well-respected scholar of the American political system and served as the Director of the Center for Congressional and Presidential Studies at AU for many years. Jim was the first person I talked to about this project and would not have undertaken it without his enthusiastic support and guidance.

Several former colleagues from Natsource made specific contributions that I want to acknowledge. First, Jack Cogen, the founder and the CEO of Natsource. He also encouraged me to write this book and provided invaluable advice and comments throughout, particularly on all things regarding the carbon markets. Andy O'Connor, the General Counsel of Natsource, who kept me out of trouble while I was with the company and continued to do so during the project. He provided invaluable advice, particularly on the sections of the book about the company and the transactions. Michael Intrator helped me in thinking about the carbon markets. Rina Cerrato reviewed my assessment of the Clean Development Mechanism and offered valuable comments and research assistance. Another former colleague, Ben Feldman provided his inputs and was particularly helpful with the details of the Atlantic Methanol Production Company (AMPCO) Clean Development Mechanism (CDM) project that is described. Rob Youngman and Dirk Forrister also supported my efforts.

Jae Edmonds is one of the world's great climate modelers and the best energy analyst I have worked with during my 25 years on the climate change issue. As usual, he was selfless in giving his time to discuss data-related issues with me, explaining differences in various data sets and reviewing some of my interpretation of them. Finally, Dan Isaac provided me with invaluable research assistance. I am grateful to all these people for their help and support.

I want to acknowledge the support of other professional colleagues over the past 25 years. There are many, particularly those I worked with on the issue of climate change. Thanks to everyone at Natsource. It was a great ride for nearly 15 years and a lot of fun to be one of the companies that pioneered the carbon markets. The early days were crazy and exhausting but I would not have changed anything. Many talented and dedicated people worked at the company and were responsible for its success.

Contents

Contents

List of Abbreviations

AB 32	California Global Warming Solutions Act
ACES	American Clean Energy and Security Act
AEO	Annual Energy Outlook
AIJ	Activities Implemented Jointly
AR	Assessment Report (IPCC)
ARB	Air Resources Board
ARRA	American Recovery and Reinvestment Act
BAU	Business as Usual
BSA	Burden Sharing Agreement
CAAA	Clean Air Act Amendments of 1990
CAFÉ	Corporate Average Fuel Economy
CAP	Climate Action Plan
CARB	California Air Resources Board
CCAP	US Climate Change Action Plan
CCS	Carbon Capture and Storage
CCTI	Climate Change Technology Initiative
CDM	Clean Development Mechanism
CEQ	Council on Environmental Quality
CERs	Certified Emission Reductions
CERUPT	Certified Emissions Reduction Unit Procurement Tender
CO_2	Carbon Dioxide
COO	Chief Operating Officer
COP	Conference of the Parties

CPP	Clean Power Plan
CPs	Complementary Policies
CQ	Congressional Quarterly
DNA	Designated National Authority
DOA	Department of Agriculture (US)
DOE	Department of Energy (US)
DOEs	Designated Operational Entities
DRM	Delivery Risk Model
EB	Executive Board
EC	European Commission
EEA	European Environment Agency
EED	Energy Efficiency Directive (EU)
EIA	Energy Information Administration
EIT	Economies in Transition
EPA	Environmental Protection Agency (US)
EPACT	Energy Policy Act
ERU	Emission Reduction Units
ERUPT	Emission Reduction Unit Procurement Tender
EU	European Union
EU ETS	European Union Emissions Trading Scheme
EUAs	European Union Allowances
GG-CAP	Greenhouse Gas Credit Aggregation Pool
GHG	Greenhouse Gas
GWP	Global Warming Potential
HFC	Hydrofluorocarbon
HG	Mercury
IEA	International Energy Agency
INDC	Intended Nationally Determined Contribution
IPCC	Intergovernmental Panel on Climate Change
JI	Joint Implementation
KP	Kyoto Protocol
LC	Letter of Credit
LCV	League of Conservation Voters
LOA	Letter of Approval
LYPC	PetroChina Limited Liaoyang Petrochemical Company
MSR	Market Stability Reserve
$MtCO_2e$	Million Tons CO_2 Equivalent
N2O	Nitrous Oxide

NAP	National Allocation Plan
NATCAP	Natsource Carbon Asset Pool
NDC	Nationally Determined Contributions
NJ	Natsource Japan
NOx	Oxides of Nitrogen
OECD	Organization for Economic Cooperation and Development
OEP	Office of Environmental Policy
PCF	World Bank's Prototype Carbon Fund
PDD	Project Design Document
PERT	Canada's Pilot Emission Reduction Trading Program
PPMV	Parts Per Million Volume
PPs	Project Participants
PUC	Public Utility Commission
RD&D	Research Development and Demonstration
RECs	Renewable Energy Credits
RED	Renewable Energy Directive (EU)
RGGI	Regional Greenhouse Gas Initiative (US)
RPS	Renewable Portfolio Standards
SD	Sustainable Development
SLCPs	Short Lived Climate Pollutants
SNAP	Significant New Alternatives Policy Program
SO_2	Sulfur Dioxide
UCF	Umbrella Carbon Facility
UNFCC	United Nations Framework Convention on Climate Change
USIJI	United States Initiative on Joint Implementation
WB	World Bank

1

Introduction

The Rationale for this Book

This book is an outgrowth of my attempt, over 25 years in senior positions in government and business, to create and use environmental markets to reduce emissions of conventional air pollutants and GHGs that cause climate change. My goal is to draw on this experience to contribute to the continuing efforts to develop effective, enduring responses to the critical issue of global climate change. I attempt to do this by describing the key policies, analyzing their results, and using these lessons to propose a path forward. A brief word on what the book is not. It is not meant to be an exhaustive review of every climate decision and policy from the last 25 years or to focus on issues such as adaptation. Others are better equipped to do that.

Those who have dedicated their careers to creating policy responses to climate change and participating in the markets understand how challenging the effort has been. There have been many successes and failures. However, given the increases in both emissions and concentrations of GHGs in the atmosphere and the resultant impacts and climate-related

© The Author(s) 2016
R.H. Rosenzweig, *Global Climate Change Policy and Carbon Markets*, Energy, Climate and the Environment,
DOI 10.1057/978-1-137-56051-3_1

risks,[1] it is fair to say that in the first generation of climate change policy-making, which I generally refer to as 'climate change 1.0', failures outweigh successes. Climate change 1.0 ended with the defeat of GHG cap-and-trade legislation in the US and generally with the initiation of negotiations for a successor treaty to the KP[2] at the international level.

The new era of policy-making, 'climate change 2.0', overlapped with 1.0 in 2009 with the advent of President Obama's policies including the first proposed regulation in the US designed to reduce GHG emissions from the transportation sector[3] and the attempt to negotiate a new international treaty[4] at the international level. It was expedited in 2013 in the US with the release of President Obama's Climate Action Plan[5] and internationally with an agreement reached at the Conference of the Parties (COP) 17[6] meeting to conclude a successor agreement to the KP in 2015. My references to international-level policy throughout the book are to the United Nations Framework Convention on Climate Change (UNFCCC),[7] the KP[8] and its market-based mechanisms, the recently concluded Paris Agreement[9] and many of the key decisions taken in the negotiations conducted under the authority of the United Nations.

My thesis is simple. The primary policy responses in climate change 1.0 failed because they were overly ambitious, complex, inflexible, and, in

---

[1] Synthesis Report Summary for Policymakers, IPCC, 2014: Climate Change 2014: Synthesis Report. Contribution of Working Groups I, II, and III to the Fifth Assessment Report of the Intergovernmental Panel on Climate Change (Core Writing Team, R.K. Pachauri and L.A. Meyer [eds.]). IPCC, Geneva, Switzerland, 151 P.

[2] *Kyoto Protocol to the United Nations Framework Convention on Climate Change.* United Nations. 1998.

[3] This is a reference to a proposed regulation which imposed the first GHG standards on light duty vehicles in the US.

[4] *Report of the Conference of the Parties on its fifteenth session, held in Copenhagen from 7 to 19 December 2009.* United Nations. 2009.

[5] Executive Office of the President. *The President's Climate Action Plan.* The White House. 2013.

[6] *Report of the Conference of the Parties on its seventeenth session, held in Durban from 28 November to 11 December 2011, Establishment of an* Ad Hoc *Working Group on the Durban Platform for Enhanced Action.* Decision 1/CP.17. United Nations. 2012.

[7] United Nations. *United Nations Framework Convention on Climate Change.* 1992.

[8] *Kyoto Protocol.* 1998.

[9] *The Paris Agreement to the United Nations Framework Convention on Climate Change.* United Nations. 2015.

part, top-down in nature. This is particularly the case given that these were the first attempts to address climate change, which is a century scale issue. In addition, the KP and its mechanisms were administered by the UN bureaucracy as overseen by nearly 200 countries. Decision-making in this process is cumbersome at best, making it extremely difficult to learn and adapt to new information. This is critical to successfully addressing any public policy issue and climate change in particular, because of its multiple dimensions and continually increased understanding of its causes.

Much has been learned from these missteps. The failures in 1.0 have significantly influenced emerging policy-making in 2.0. For the most part, the new efforts underway in the US to achieve GHG emission reduction targets are more bottom-up and targeted in nature and consist of less ambitious measures[10] than a broad-based tax or an economy-wide cap-and-trade program as proposed in 1.0.[11] Similarly, the international agreement recently concluded in Paris[12] at the twenty-first COP is in total contrast to the KP. The approach to reducing GHG emissions to achieve climate policy objectives is bottom-up in nature, flexible and supported by top-down elements. Hopefully, it will work better in slowing global GHG emissions.

Although US and international climate change policies in 2.0 will not be as ambitious as the ones that were considered and adopted in 1.0, they will likely be more successful. The book reviews the policies proposed in the US and adopted at the international level in 1.0, assesses why they failed, and describes how they influenced ongoing policy development. It reviews emerging trends in the new era of policy-making and propose a series of 'modest', targeted policies that I believe can be effective in reducing GHG emissions and controlling costs. Success is essential to building public confidence and creating the political conditions necessary to develop more ambitious actions that will be required in the future, something that is imperative in today's fractured and often dysfunctional political environment. To borrow an analogy from baseball, it is time to play a

---

[10] The US strategy, and the policies which comprise it will be described in later chapters of the book.
[11] The policies most identified with climate change 1.0 in the US are the BTU tax proposed by President Bill Clinton in 1993 and the American Clean Energy and Security Act which passed the House of Representative in 2009 and died in the US Senate. These initiatives will be described in subsequent chapters.
[12] *The Paris Agreement.* 2015.

small ball. Policy-makers should attempt to hit a lot of singles and doubles in trying to achieve large-scale reductions in GHG emissions. Singles and doubles, in this context, modest initiatives, put a lot of runs on the board. We need to avoid the temptation to swing for the fences and hit home runs in 2.0. The overreach in 1.0 was a primary cause of its failure.

A review of the first generation of climate change policy and recommendations for 2.0 would be incomplete without a sober assessment of the performance of the carbon markets that were the cornerstone in 1.0. With great fanfare, the KP attempted to create a single, integrated global market to assist developed countries achieve GHG emissions reduction obligations at the lowest cost. Similarly, the EU Emissions Trading Scheme[13] (EU ETS) was the primary element of its strategy to comply with its KP targets and was linked to the Kyoto market. It remains a cornerstone of EU climate policy to achieve long-term GHG emission reduction targets. The US failed in its attempt to create a carbon market by developing an economy-wide GHG cap-and-trade system in the first few years of the Obama Administration.

The carbon markets were and continue to be a significant source of controversy. In my view, advocates oversell markets' potential benefits while detractors minimize them. I have not found many dispassionate reviews of their actual performance. Therefore, one of the book's primary emphases is to undertake and provide such a review of the performance of the KP markets, with an emphasis of the CDM[14] and the EU ETS. A clear look at their shortcomings and successes, and the reasons for such, is essential to understand what carbon markets can realistically deliver in the future. This is critical, given that approximately 60 trading systems or taxes have already been implemented or are under development at the national and subnational levels and a new mechanism was incorporated in the Paris Agreement.[15,16] As such, markets will continue to be a prominent component of the climate policy portfolio in 2.0.

---

[13] European Commission. *The EU Emissions Trading Scheme.* European Commission. doi: 10.2834/55480. 2013.

[14] *Kyoto Protocol. Article 12.* 1998.

[15] Kossoy, A., G. Peszko, K. Oppermann, N. Prytz, N. Klein, K. Blok, *State and Trends of Carbon Pricing* 2015 (September), by World Bank, Washington, DC.

[16] *The Paris Agreement. Article 6.* 2015.

I was a strong believer that carbon markets provided the best hope of achieving climate policy objectives at the lowest cost. My support was based primarily on my experience with the US acid rain trading program (the world's first large-scale market created to solve an environmental problem) included in the Clean Air Act Amendments (CAAA) of 1990. However, it was strengthened by the characteristics of GHG emissions and economics. First, the nature of climate change is such that reductions achieved anywhere in the world would benefit the global climate equally. And second, the large disparities in reduction costs around the globe provided powerful cost-saving opportunities for trade. Designed correctly, carbon markets could help drive down the cost of achieving GHG emission reductions, provide incentives for additional reductions, and stimulate innovation. In the US, markets also provide the possibility of moving past the contentious and inevitable debate over the use of taxes to address climate change as they had with acid rain.

After stubbornly denying it for many years, I reluctantly concluded that the initial vision of carbon markets playing the central role in GHG emissions mitigation and mobilizing large volumes of capital necessary to combat climate change would not become a reality. This was a difficult conclusion for me to reach. Markets will continue to a play an important role in the effort to address climate change; however, other approaches will also play significant roles. Policy-makers and affected parties need to move past the contentious debates of trade versus taxes versus regulation. They all have a role to play and we need all of them. Each nation should implement policies based on their circumstances and policy-making traditions.

My conclusions regarding the role that markets have played in 1.0 and their best use in 2.0 result from my experiences in government as the Chief of Staff at the US DOE from 1993 to 1996 and in the private sector as the Managing Director and COO from 2000 to 2013 of Natsource, a leading company in the formative years of the carbon markets.

At DOE, I participated in the development of the first project-based market mechanism designed to reduce GHG emissions. The mechanism, the United States Initiative on Joint Implementation (USIJI), was a pilot program included in the first US climate change action plan (CCAP) developed in 1993.[17] It was an outgrowth of the Joint Implementation (JI) concept that was included in the UNFCCC.[18] USIJI, along with other pilot programs including activities implemented jointly (AIJ),[19] which was created by the international community, were the forerunners of the CDM, included as Article 12 in the KP, which became an important and controversial component of the global carbon market. The USIJI was included in the CCAP in recognition of the global opportunities to reduce GHG emissions where they were the cheapest and for firms to gain experience investing in emission reduction projects outside of the US. It would be the first step in attempting to determine if such programs could work and in motivating the private sector to operationalize such mechanisms. Following my departure from government, and prior to joining Natsource, I worked with several large utilities and energy companies during the Kyoto negotiations and to formulate response strategies once it had been agreed.

In recognition of the Bush Administration's decision to avoid the issue of climate change, I departed the familiar policy world to gain commercial experience and with environmental markets. I joined Natsource in 2000 to create a research business to work with the private sector on climate change. I became the COO in 2005—the year the KP took effect and when the company launched the world's largest private sector carbon fund. According to an independent research, Natsource was the largest buyer of contracted carbon credits created by the KP mechanisms on behalf of its investors through 2007 on a risk-adjusted basis.[20] The company closed in 2014.

My hope for the carbon markets in climate change 1.0 was not realized for many reasons. Among the most important are the artificial nature of environmental markets and that the people who design them

---

[17] Clinton, President W.J., Vice President A. Gore Jr. *The Climate Change Action Plan.* Executive Office of the President. 1993. PP. 26–27.

[18] *UNFCCC.* Article 4.2. (a). 1992.

[19] *Report Of The Conference Of The Parties On Its First Session, Held At Berlin From 28 March To 7 April 1995, Activities implemented jointly under the pilot phase.* Decision 5/CP.1. United Nations. 1995.

[20] Rosenzweig, R. Natsource Recognized as World's Largest Purchaser of Carbon Credits by Leading Investor Research Firm. (Press Release) 6 March 2008.

frequently lack commercial and financial expertise. Unlike natural markets, no firm would require a GHG emissions offset or permit/allowance unless Government required them to comply with an emissions limitation. Governments establish the supplies of compliance instruments in these markets and attempt to set demand, although other dynamics intervene in these efforts, particularly on the demand-side. This leads to design elements that adversely affect the markets' performance. These issues will be described in Chaps. 3 and 4 assessing the CDM and the EU ETS.

In addition, the EU ETS and the Kyoto mechanisms share a characteristic, which is common to top-down systems, that greatly affected their performance in 1.0. This is the inability of governments to respond to and learn from external events and adapt to new information in a timely fashion. For example, the economic recession, which took hold in 2008, and the energy policies in place at the EU level, which operated alongside the EU ETS, contributed to a massive supply and demand imbalance beginning in 2009 that continues today. And although the market enjoyed some successes, the EUs inability to respond contributed to the market's uneven performance and volatility since their inception.

The issues regarding the artificial nature of the market, program design, and external dynamics will continue to impact market performance at the international level. And although the US never adopted a national carbon market, and will not for the foreseeable future, the attempt to pass legislation that would have created a market for GHG emissions following the election of President Obama was a failure by any measure. Its demise was caused by several substantive and political reasons that will be the subject of discussion in Chap. 5. I am confident that the market would not have worked as intended. These conclusions regarding carbon markets are what I reluctantly took away from Climate Change 1.0. Supporters of the KP model believed it would provide Parties with an incentive to develop domestic cap-and-trade systems that would link to the global market. Many believed that this approach provided the best hope to achieve climate policy objectives at the lowest cost. It did not work out as they had hoped.

Carbon markets have an important role to play in the policy portfolio in climate change 2.0 at the national and subnational levels—but they are not 'the' answer as many had thought. Taxes and regulation will also play prominent roles in future climate change policy-making. The argument as to which is the best approach needs to stop. The reality of the 60 diverse, bottom-up programs in existence and under development is quite different from the KP's top-down approach to creating a global market. With the exception of the EU ETS and China's effort to create a carbon market, they are less ambitious. However, their performance will continue to be affected by the dynamics cited above. For example, the EU ETS operates alongside many other energy policies and measures, many of them regulatory in nature, designed to achieve ambitious goals for renewable energy[21] and energy efficiency.[22] Similarly, California's cap-and-trade program operates alongside many other measures called complementary policies (CPs).[23] Other jurisdictions are using similar policy models in their response to climate change. These programs often compete with the market's primary objective of achieving GHG reductions at the lowest cost. The interaction between the market-based systems and regulatory approaches in the emerging era of policy-making will have a significant impact on the magnitude of GHG reductions that are achieved and their costs. To inform future policy-making efforts, more research is required to gain a greater understanding of the interactions between these policy approaches.

Because of continuing interest in using market-based approaches to achieve climate policy objectives, the book will briefly describe the evolution of these approaches and provide recommendations regarding their future role in the policy portfolio. My experience in government

---

[21] Directive 2009/28/EC of the European Parliament and of the Council of 23 April 2009 on the promotion of the use of energy from renewable sources and amending and subsequently repealing Directives 2001/77/EC and 2003/30/EC. Brussels, European Parliament, and Council.

[22] Directive 2012/27/EU of the European Parliament and of the Council of 25 October 2012 on energy efficiency, amending Directives 2009/125/EC and 2010/30/EU and repealing Directives 2004/8/EC and 2006/32/EC. Brussels, European Parliament, and Council.

[23] California Air Resources Board. *Climate Change Scoping Plan, A Framework for Change Pursuant to AB 32, The Global Warming Solutions Act.* 2008.

and participation in the market provides practical insight into these important issues.

Overview

This book is organized around two generations of climate change policy that are primarily distinguished by differences in their approaches and ambition.

Chapters 2, 3, 4 and 5 focus on what I refer to as climate change 1.0. Collectively, this is a reference to domestic and international policies and carbon markets from 1993 through 2012. To learn from these experiences, the main emphasis is on the reasons for the US's failure to adopt important climate change policies including the tax on the British Thermal Unit (BTU) content of energy and GHG cap-and-trade legislation. At the international level, the emphasis is on the major policies and markets including the EU ETS, the CDM, and the KP.

Managing a company at the dawn of the carbon markets was a great personal and professional challenge; it was both exhilarating and exhausting. In an attempt to make the markets less abstract for the reader and show firsthand how they operate, Chaps. 3 and 4 describe Natsource's business strategy to participate in the markets, some of the cutting-edge transactions the company participated in, and the forces that contributed to the closing of the company. I am hopeful that these real-world examples will be entertaining, but most importantly illustrate the interaction between policies and markets and how companies participate in them.

Chapter 6 describes the emergence of policies that are defining climate change 2.0 in the US and at the international level. Collectively, this is a group of more targeted bottom-up policies that emerged while the first generation was drawing to a close. They include President Obama's Climate Change Action Plan and the Paris agreement. Chapter 7 provides recommendations for future policy in the US and internationally.

2

Climate Change Policies of the Clinton Administration

Domestic Policy

The environmental community was excited for the Clinton Administration to take office following 12 years of Republican administrations it viewed as hostile to the environment. President George H.W. Bush had advocated for and signed the CAAA of 1990 into law and his administration had negotiated the UNFCCC, the international community's first attempt to develop an international framework to address the issue of climate change. It included the non-binding aim of returning GHG emissions to 1990 levels.[1] However, the environmental community did not believe that the voluntary emissions reduction goals included in the convention were up to the task. After 12 years in the wilderness, it was looking forward to working with the new Administration and Democratic majorities in both houses of Congress to advance its agenda, and one of its primary emphases was on climate change. During the campaign, candidate Clinton had pledged to 'limit US carbon dioxide emissions

---

[1] United Nations. *United Nations Framework Convention on Climate Change.* Article 4 2. (b). 1992.

© The Author(s) 2016
R.H. Rosenzweig, *Global Climate Change Policy and Carbon Markets*, Energy, Climate and the Environment,
DOI 10.1057/978-1-137-56051-3_2

to 1990 levels by the year 2000'.[2] The environmental community was particularly enthused with the selection of Al Gore as vice president. He was a committed environmentalist, a recognized expert on the subject of climate change, and had authored a book on environmental issues, *Earth in the Balance*,[3] in 1992.

Looking back nearly a quarter of a century, those supportive of taking action to address climate change might not be so happy with the results.

Fiscal and Climate Policy?

The first order of business for the Clinton Administration was to stimulate economic growth and tackle the budget deficit. I started at DOE on 20 January 1993, the same day of President Clinton's inauguration and Secretary Hazel O'Leary's confirmation as the Energy Secretary by the US Senate. Almost immediately, an interagency process was initiated to formally analyze energy tax options for consideration in the president's budget proposals. DOE was key to this effort, given its analytical capabilities, understanding of the issues, and relationships with the industries that would be most affected by such proposals. Several energy tax options, including a motor fuels tax, oil import fee, carbon tax, BTU tax, and ad valorem taxes, were assessed for their macro-economic impacts on the Gross Domestic Product (GDP), employment, revenue-generation, and distributional impacts on different income groups, and regions. They were also assessed for their impacts on energy prices, oil markets, energy production, and energy consumption. And, they were assessed for their impact on reducing GHG emissions.[4]

The BTU tax was included in the budget deficit proposal President Clinton delivered in his first speech to Congress on 17 February 1993. The groundwork for an energy tax had been laid prior to President Clinton taking office. The Environment, Energy, and Natural Resource Options Book prepared by the President-elect's transition team in December of

---

[2] Clinton, Governor B., Senator A. Gore (1992) *Putting People First: How We All Can Change America* (Three Rivers Press). P. 97.

[3] Gore, A. (1992) *Earth in the Balance: Ecology and the Human Spirit* (Emmaus, PA; Rodale, Inc.).

[4] Office of Domestic and International Energy Policy. *Briefing on Energy Taxes*. US Department of Energy. This was a document that I maintained which was prepared for the Energy Secretary. 1993.

1992 evaluated the impacts of a carbon tax, a BTU tax, and a gasoline tax under the issue of deficit reduction.[5] The arguments in favor of the BTU tax included in the briefing book were similar to those used by President Clinton in proposing the tax before Congress.[6]

President Clinton's US$500 billion deficit reduction package was split equally between spending cuts and tax increases. The BTU tax was the largest revenue component; it was estimated to raise approximately US$70 billion of the US$250 billion in revenues that the president was seeking.[7] Importantly, it was also estimated that the BTU tax would reduce carbon emissions by 43 million MT in 2000.[8]

The president recommended a BTU tax on the heat content of energy, saying it 'combats pollution, promotes energy efficiency, promotes the independence, economically, of this country as well as helping to reduce the debt, and because it does not discriminate against any area'.[9] Although the proposed tax was sold primarily for its ability to raise revenue, the reference to combating pollution was to GHG emissions. In effect, even though the speech did not mention climate change, the BTU tax was the Administration's first policy proposal to address climate change. It kicked off climate change 1.0.

President Clinton Makes First GHG Emissions Commitment

There was great anticipation in advance of President Clinton's first earth day speech at the Botanical Gardens in Washington, DC on 21 April 1993.

---

[5] Environment, Energy and Natural Resource Options book. This was a document prepared for the Clinton administration. 1993. PP. 72–83.

[6] Environment Options book, 1993. PP. 77–78.

[7] William J. Clinton: 'Address Before a Joint Session of Congress on Administration Goals', February 17, 1993. Online by Gerhard Peters and John T. Woolley, *The American Presidency Project*. http://www.presidency.ucsb.edu/ws/?pid=47232.

[8] *Briefing on Energy Taxes*. 1993.

[9] William J. Clinton: 'Address Before a Joint Session of Congress on Administration Goals', February 17, 1993. Online by Gerhard Peters and John T. Woolley, *The American Presidency Project*. http://www.presidency.ucsb.edu/ws/?pid=47232.

In his speech, President Clinton said, '[w]e also must take the lead in addressing the challenge of global warming that could make our planet and its climate less hospitable and more hostile to human life. Today, I reaffirm my personal and announce our Nation's commitment to reducing our emissions of greenhouse gases to their 1990 levels by the year 2000...I am instructing my administration to produce a cost-effective plan by August that can continue the trend of reduced emissions. This must be a clarion call, not for more bureaucracy or regulation or unnecessary costs but, instead, for American ingenuity and creativity, to produce the best and most energy-efficient technology.'[10]

There was nothing particularly far-reaching about this commitment. Anything less would have been seen by the Administration's supporters as a major retrenchment. It generally codified the short-term GHG emissions goal incorporated in the UNFCCC for a group of developed countries. However, two months after the budget address it was clear that the BTU tax was in for a rough ride.

At the same time, that DOE was evaluating energy taxes for the deficit reduction effort, it was also participating in an interagency process on the climate change issue. This discussion was in full swing prior to President Clinton's Earth Day speech at the Botanic Gardens. The process was chaired by the National Security Council and the White House Council on Environmental Quality (CEQ). Because President Clinton had committed to reduce GHG emissions during the recently completed presidential campaign, the issue at play was 'what policy framework would be included in the president's plan to achieve the goal?' Would it be a flexible, market-based policy such as a GHG cap-and-trade system that had grown in popularity based on the acid rain trading program incorporated in the CAAA of 1990, taxes, or traditional regulation?

As is typical of these types of interagency processes, conflict quickly arose the between the Administration's environmental and economic teams. Senior leadership of the Department of the Treasury was focused on cost, and they asked advocates for the GHG emissions reduction goal what they would tell President Clinton if he questioned the cost to

---

[10] William J. Clinton: 'Remarks on Earth Day,' April 21, 1993. Online by Gerhard Peters and John T. Woolley, *The American Presidency Project*. http://www.presidency.ucsb.edu/ws/?pid=46460.

achieve the goal. There was no specific response. Supporters of the decision argued that a cost-effective policy would allow for (i) reductions of all GHGs, and not just carbon dioxide (CO_2); (ii) crediting reductions achieved by biologic or terrestrial sequestration, as opposed to reducing GHGs solely from emissions sources; and, (iii) using market-based approaches in lieu of traditional regulation. In the early days, this policy framework was known as the 'comprehensive approach'.

I represented DOE on behalf of Secretary O'Leary in the interagency discussions. At the time, many of the environmental advocates within the Administration viewed DOE skeptically. They saw DOE as having served as the right flank on climate policy in the George H. W. Bush Administration, essentially doing the bidding of those in industry that opposed any action to reduce GHG emissions. Our goal was to position DOE in the middle of the debate and to get a market-based policy framework in place to achieve GHG reductions at the lowest possible cost. In our view, this was the responsible position and would also enable DOE to represent its energy constituency effectively in the emerging debate. We were able to position DOE as a supporter of the policy decision while advocating forcefully for a market-based framework. At the end of the day, the Administration emphasized its preference for the comprehensive approach to meet GHG emissions goals. The key would be in developing the cost-effective plan the president referred to in his Earth Day speech.

The BTU Tax Lives a Short Life

Although it was not advertised as a climate change initiative, if adopted, the BTU tax could have had a significant impact on future domestic climate change policy, possibly eliminating the need for other policies, such as proposed cap-and-trade programs that are discussed later in the book. It was not only defeated, but the process of defeat increased the polarization of the climate change issue and greatly influenced future debates, as is discussed later. In that regard, its impacts can even be felt today.

Why Environmental Taxes?

Many academics and environmentalists had advocated for years that imposing taxes on emissions was the most efficient way to achieve reductions of harmful pollutants. The argument is that emissions created by energy production and use cause air and water pollution resulting in health and other environmental problems that, in turn, impose costs on the entire society. These are known as 'environmental externalities'.[11] Typically, emissions are free for the polluter and the costs are external to them. The theory is that a tax on energy, or the emissions they cause, imposes costs on the polluter, providing a financial incentive for them to reduce emissions. Thus, the tax would result in the firm internalizing the cost while also providing a revenue source for the government. Taken one step further, some believe that the imposition of taxes on the activities that cause environmental problems such as climate change could provide an impetus to overhaul the tax code, providing benefits to the overall economy. In theory, taxes on 'bad things' like pollution, such as GHG emissions, would raise sufficient revenue, enabling the reduction of taxes on 'good things', such as on income and capital.

I have learned never to say never; but I do not believe this theory will bear out in the US anytime soon—or at least in my lifetime—as it relates to climate change or any other large-scale environmental challenge. Americans like their energy cheap, and its low-cost has long benefitted US industry. Taxes on energy would increase its cost, potentially disadvantaging the industry. Advocates argue that the tax would provide an incentive for industry to become more efficient, eventually leading to reduced energy use, lower costs, and increased competitiveness. Energy taxes are also regressive. Low-income households spend a higher percentage of their income on energy than high-income households. Energy taxes would exacerbate this problem. Advocates counter by saying the energy tax could be designed to address this problem. However, the substantive issues raised by energy taxes do not even consider the

---

[11] United Nations. *Glossary of Environmental Statistics*. Report number: ST/ESA/STAT/SER.F/67, 1997.

acrimonious discourse that occurs in the US Congress over any proposal to increase revenues.

I was biased against taxes as this process began. Academics categorize taxes as market incentives, similar to the markets created by cap-and-trade systems for the purpose of reducing GHG emissions. Both mechanisms put a price on emissions. Taxes provide certainty with respect to costs, while cap-and-trade systems provide certainty with respect to the level of emissions. There is an exhaustive literature on the similarities and differences in price-based market instruments such as taxes, and quantity-based market instruments such as cap-and-trade. I strongly preferred emission markets for several reasons. The preeminent one was that I believed an emissions market created by a well-designed cap-and-trade program would enable private firms to make better decisions to innovate and reduce their GHG emissions than the government would make by spending the revenues generated by a tax. My experience in government convinced me of this.

The Politics of BTU

The BTU tax was a disaster for the Clinton Administration, and, as it turns out, for climate change policy in the long run. The phrase 'I've been BTUed' was coined as a result of the fallout from congressional debate on the tax in which several members of congress felt they had gotten burned by voting for it. Interestingly, there were clear analogies between the debates over the BTU tax in 1993 and over cap-and-trade policy 16 years later in 2009. The arguments and language used by opponents of the tax and cap-and-trade were nearly identical. This is discussed in Chap. 5.

Attempts to pass the BTU tax quickly exposed fissures between the Administration's political and policy objectives. Almost immediately, political officials expressed their concern that the tax would be visible to the public. Higher prices for electricity and gas would show up on customer's monthly bills. Prices for other energy commodities would also go up. Some were concerned that voters would blame the Administration's BTU tax for this increase. Taken one step further, the resulting backlash could adversely affect the president's reelection campaign and potentially damage the reelection prospects of congressional supporters. There was also concern about the potential impact on Vice President Gore's expected campaign for the presidency. So, although the policy purpose of the tax was to modify behavior by raising energy prices, the politics argued for hiding those higher prices from the consumer.

In an attempt to address the political concerns, some argued for imposing the tax at varying points in the energy production process instead of placing it on the end user. Although this approach attempted to hide the tax from consumers, it conflicted with the policy objectives. If the tax was going to be successful in achieving its environmental and efficiency objectives, consumers needed to see higher prices and respond by reducing their energy use. This geeky design issue generated intense and lengthy internal debate within the Administration. In truth, it did not matter where the tax would be imposed. It could not be hidden regardless of the intent to do so. The internal debate within the Administration ignored the influence that state public utility commissions (PUCs) would play in the process. For example, in their desire not to be blamed for higher energy bills, state PUCs would require electric and gas companies to put the tax on their bill. Given this likely outcome, it made sense to focus on the policy objective and put the tax on the end user.

The attempt to hide the tax had many unintended consequences. One of the most important of which was to turn the gas industry in to a vocal opponent of the BTU tax. A cornerstone of President Clinton's energy policy was to stimulate increased gas use, and the tax was expected to facilitate this. In theory, the tax could have created significant support from the gas industry, creating a split with other sectors of the industry. However, imposing the BTU tax at some point in the production process pitted the various sectors (producers, transporters, and distributors) of the gas industry against each other. In the end, they could agree on nothing but to oppose the tax, and the industry was generally united in its opposition. These political dynamics were not considered in the internal fight over the tax design.

The politics of the BTU tax were ugly. Opponents quickly mobilized and coalitions were established. Their message was simple and it hit home: the BTU tax would increase the costs of goods and services and it would kill jobs by eliminating the US manufacturing sectors' traditional advantage of low-cost energy.[12] Some argued that the tax would kill 600,000 jobs and it was labeled a job destroyer.[13] In contrast, communicating simple messages to the public on the tax benefits was much more challenging. As is described in Chap. 5, this same dynamic played out in the debate over cap-and-trade.

At the DOE, our outreach efforts to industry and other important constituencies were not going well. The briefings held by the Administration in the White House for the coal, oil, natural gas, and electric utility industries were not pleasant given the strong industry opposition. And, of course, the energy industry was relentless in lobbying sympathetic Members of Congress, which helped reinforce strong and widespread opposition to the tax on Capitol Hill. Representatives from affected states were also unhappy with the BTU tax. For example, I recall a briefing that I did in the Old Executive Office Building for State legislators from energy-producing states. I stood-in for Mac McLarty, the White House Chief of Staff and former CEO of a gas company, who had a conflicting meeting. This was a bad start and the briefing only got worse. A DOE official was not an adequate substitute for the Chief of Staff to the president who was viewed as an industry ally, so attendees were not happy from the outset. Because the president's budget had not yet been sent to Capitol Hill, I could not say much about the details of the BTU tax and the briefing went downhill. Attendees assumed I was evading their questions. I recall one energy trade press reporter labeling me as a 'slightly arrogant twerp' in his story. In consolation, my former business partner delighted in telling me I was arrogant, but not a twerp.

Congress was the most important constituency because it would be the ultimate decision-maker on the BTU tax. Our attempts to win them over were not going much better. Conservative Democrats from energy-producing states were generally opposed to higher taxes and were particularly opposed to the BTU tax because adverse impacts on energy production would cut jobs and revenues in their districts and states.

---

[12] Erlander, D. (1994) 'The BTU Tax Experience: What Happened and Why It Happened', *12 Pace Envtl. L. Rev.* 12(1): 173–184.

[13] D. Erlander (1994) 'The BTU Tax Experience', P. 179.

Yet, they did not want to oppose a major initiative from a newly elected president who was from their own region and party, and who had a reputation for being a moderate. They were upset with the answers they were getting to their questions and were convinced the tax was the product of Vice President Gore, a figure who drew skepticism from these Democrats based on his environmental views.

Discussions with supposed allies grew increasingly contentious over both substance and form. For example, in one briefing for members and staff of the House Energy and Commerce Committee, a senior southern Democrat challenged a White House staffer to identify their role. She explained that she was an adviser to the president. To which the Congressman rejoined, 'Well, I would advise you to advise the President to call me.' And this was during the high point in our efforts to sell the BTU tax. Senior Members of Congress prefer to talk to Cabinet Secretaries and Presidents, not their staff.

The House of Representatives had the first vote on the president's budget. To pass the bill, the Administration needed the votes of conservative Southern Democrats like Billy Tauzin (D-LA), an influential Member of Congress on energy policy issues. On the night of the final vote, several Southern Democratic House members were clustered together. In an attempt to secure their support for the bill, the Administration committed to fix problems with the tax after passage and prior to it being debated in the US Senate. In this scenario, House members would get credit for the changes. Based on this and other commitments, the Administration secured key votes for the budget and it passed by a 219–213 margin on 27 May 1993.[14] Ultimately, the US Senate turned the BTU tax into a 4.3-cent gasoline tax. House members never did get credit for fixing the tax; instead they took the hard political vote and got the criticism back home. The Senators got credit for effectively eliminating the tax.

Politicians take great stock in commitments made by their colleagues in the Congress, the Executive Branch, and at all levels. Because many of the House members who supported the legislation did not receive the cover they believed the White House promised, those members felt as if

---

[14] Final Vote for Roll Call 199. 'Omnibus Budget Reconciliation Act of 1993'. Available at the Clerk of the House Web Site: http://clerk.house.gov/evs/1993/roll199.xml (27 May 1993).

they had been lied to and their chances for reelection had been compromised. House members coined the phrase 'I got BTUed' to describe their reactions to what had occurred. They vowed 'never again' and the relationships between these House members and the Administration were badly damaged. Congressman Tauzin switched parties in 1995. The budget finally became law in August of 1993 after it squeaked by the House with a 218–216[15] margin and the Senate with a 51–50 margin with Vice President Gore casting the tie-breaking vote in his role as the President of the Senate.[16]

Lasting Impacts from the BTU Debacle

In the context of climate change 1.0, the debate over the BTU tax was important even though it was defeated. Although it was advertised and sold primarily to raise revenue to reduce the budget deficit, it was also expected to reduce GHG emissions and achieve other policy objectives. Discussions were held regularly in the Administration on the tax's impact in reducing GHG emissions. Advocates wanted to get the tax in place and business wanted to kill it. They both understood that once a tax is in place, the potential exists to increase it. And, a broad-based tax on energy had the potential to raise massive amounts of revenue. This is one of the key reasons why the business community opposed the BTU tax so strongly. So, although the tax was not as important as the attempt to pass a cap-and-trade bill that is discussed later, it would have been a serious climate policy initiative and the ill effects of the process are still being felt in the debate on US climate change policy.

From my personal vantage point, the tension between the politics and substance surrounding the BTU tax debate shaped my views regarding the potential for taxes to be used to address climate change in future US policy-making efforts. Taxes had been considered as one of the primary

---

[15] Final Vote for Roll Call 406. 'Omnibus Budget Reconciliation Act of 1993'. Available at the Clerk of the House Web Site: http://clerk.house.gov/evs/1993/roll406.xml. (5 August 1993).
[16] On the Conference Report to H.R. 2264. 'Omnibus Budget Reconciliation Act of 1993'. Available at: United States Senate Web Site: http://www.senate.gov/legislative/LIS/roll_call_lists/roll_call_vote_cfm.cfm?congress=103&session=1&vote=00247 (6 August 1993).

policies available to achieve large-scale reductions in GHG emissions. One of the political impacts of the debate on the BTU tax and its demise was to eliminate the potential for taxes to be used to reduce GHG emissions for the foreseeable future. Even today, it remains unlikely taxes will be in the US climate policy mix at the federal level. In addition, the deliberations on the BTU tax debate set the tone for climate change policy in the US during the rest of 1.0, and it was not pretty. It poisoned an already contentious debate. The missteps on the tax helped strengthen an already conservative group of industries opposed to taking any action to address climate change. They poured millions of dollars in the effort to defeat the BTU tax and committed millions more in the US to discredit climate science and other policy initiatives during the Clinton Administration. During these years, advocates of climate change action were not nearly as organized, united, or well-funded as their opponents. In addition to reducing the potential for taxes and poisoning the political climate, it is unclear whether cap-and-trade would have become as prominent in climate change policy-making if the BTU tax had been enacted. It might not have been required.

Substantively, many people who support an energy tax as a mechanism to achieve climate policy objectives argue that in contrast to cap-and-trade, one of its primary virtues is its simplicity. Perhaps taxes are simple in theory, but not in practice. The BTU tax proposed by the Clinton Administration was anything but simple, and it became more complex as it moved through the legislative process. It included exemptions and provisions to address concerns raised by affected interests including hydroelectricity in the Pacific Northwest that provided power to the aluminum industry, home heating oil in the Northeast, and agricultural interests. As a rule of thumb, any tax legislation designed to raise approximately US$70 billion may start off as a simple plan but will not end up that way. A tax that would actually achieve large-scale GHG reductions would need to be much larger than the BTU tax and it would end up being as complex, if not more so, than any other policy, including cap-and-trade.

The primary lesson from this effort is that the BTU tax failed in large part because it overreached. It was too complex and controversial for what it was attempting to accomplish, particularly as an early step to reduce GHG emissions. And although not as ambitious and complex as an economy-wide cap-and-trade bill or the KP, broad-based policies that affect the entire economy, such as the BTU tax, engender strong political opposition, which in this case, could not be overcome.

The Climate Change Action Plan

On 21 October 1993, President Clinton and Vice President Gore presented the Administration's CCAP[17] to achieve the GHG emission reduction goal that had been announced on Earth Day. The plan consisted primarily of voluntary measures and industry partnerships; it would cost approximately US$2 billion over six years.[18] It was designed to improve the efficiency of commercial and residential buildings, and the industrial sector, while increasing the use of natural gas and renewable energy in the power sector. Other partnerships were announced to reduce methane leaks in natural gas pipelines and to increase the use of methane from animal manure.

The plan was generally praised by industry because of its emphasis on voluntary measures. The environmental community was lukewarm: they disliked it for its lack of ambition, but did not want to criticize the Administration too strongly. It was considered to be the best the Administration could muster at the time. Following the defeat of the BTU tax just a few months earlier, it was clear that the Congress would not entertain any mandatory initiatives to reduce GHG emissions. The political realities were obvious. It was a sober reminder in 1993 that there was little possibility to make progress on the home front in attacking climate change during the Clinton Administration. And this was at a time when the Democrats controlled both houses of Congress. I participated in the effort to develop the CCAP. Part of the way through the process, I recognized that the agencies that were responsible for administering it were using the process as a budget building exercise. If you needed to scale up your voluntary programs, more resources were required. It made sense!

---

[17] Clinton, President W.J., Vice President A. Gore Jr. *The Climate Change Action Plan*. Executive Office of the President. 1993. PP. 26–27.

[18] Clinton, President W.J., Vice President A. Gore Jr. *Climate Change Action Plan*. P. ii.

Key Elements of the CCAP

The DOE worked closely with the industry, particularly electric utilities, in developing the CCAP. After extensive consultations, we worked together to create a program called the 'climate challenge'.[19] In this effort, power generators made a commitment to achieve some level of GHG emission reductions, take actions to achieve it, and report them to the department. My colleagues and I held hundreds of calls and meetings with industry representatives to secure their commitment. Companies signed up to one of the five potential levels of commitment ranging from returning their emissions to 1990 levels in 2000 at the high end to implementing a reduction project, at the low end, or participating in one of several industry-wide initiatives that were established. The initiatives were designed to increase the annual installations of geothermal heat pumps, accelerate the introduction of electric vehicles, invest in companies developing electric and renewable technologies, and implement domestic and international forestry projects.[20] The electric utility industry also established the international utility efficiency partnership to coordinate its member's participation in pilot Joint Implementation projects.[21] An industry publication communicated that more than 600 electric utilities pledged to limit their emissions by more than 170 million metric tons of CO_2 equivalent ($MtCO_2e$) through a portfolio of actions.[22]

At the time it seemed like an important accomplishment because various elements of the power sector, particularly those that relied on coal to fuel their plants, had been among the most vocal opponents of the acid rain program included in the CAAA of 1990 and of taking action to reduce GHG emissions. Because of the climate challenge, they were actively engaged in taking actions to reduce their GHG emissions and played a constructive role in the process. The joint discussions of the climate challenge also solidified the industry's commitment to market-based mechanisms. From a PR perspective, the DOE was happy to wave

---

[19] Clinton, President W.J., Vice President A. Gore Jr. *Climate Change Action Plan.* PP. 22–23.

[20] Edison Electric Institute. *Everyone has a responsibility to protect the environment.* 1998.

[21] EEI. *Everyone has a responsibility.*

[22] EEI. *Everyone has a responsibility.*

the company's letters to the Secretary committing to participate in the program during the announcement of the plan. It made for a good picture and something positive to announce. In hindsight, the program had a limited impact on GHG emissions.

The most important and enduring component of the plan was the USIJI.[23] As indicated in Chap. 1, it was an outgrowth of the JI concept included in Article 4 2. (a) of the UNFCCC, which was focused on Annex I Parties' (these were predominantly developed countries and others defined as undergoing a transition to a market economy) commitment to mitigate climate change including '…the return by the end of the present decade to earlier levels of anthropogenic emissions of carbon dioxide and other GHGs not controlled by the Montreal Protocol'.[24] Specifically, JI authorized the Parties to 'implement such policy and measures jointly with other Parties and may assist other Parties in contributing to the achievement of the objective of the Convention and, in particular that of this subparagraph'.[25]

JI was included in the UNFCCC in recognition of the opportunities to achieve GHG emission reductions cooperatively between nations, and that they could be achieved in developing countries at costs lower than that in mature, industrialized countries like the US. In 1993, some had a vision for developed country investors to implement GHG reduction projects in developing countries and to be able to use the reductions to comply with future GHG emission reduction obligations. The mechanism was also attractive for its potential to stimulate the transfer of capital, technology, and services from developed to developing countries; this was an important objective embodied in the convention. Many were also opposed to JI at the time, believing that it would not be possible to prove that reductions achieved by projects were additional, would allow industry to buy its way out of the problem rather than reducing emissions within their own assets, and generally distrusting markets. The debate has not changed much in a quarter of a century.

---

[23] Clinton, President W.J., Vice President A. Gore Jr. *Climate Change Action Plan.* PP. 26–27.
[24] *UNFCCC.* Article 4 2. (a). United Nations. 1992.
[25] *UNFCCC.* Article 4 2. (a).

Although promising, JI was still in its infancy and far greater thought was required to put meat on its bones and expand it into an acceptable program. Typically, the best way to do this is learning by doing. This meant actual projects needed to be undertaken. The USIJI was an attempt to begin that process. It was a pilot initiative that created criteria for investors to complete such projects.[26] Overall, USIJI was influential in the international negotiations and helped usher in an international pilot initiative.

Summary

The BTU tax and the CCAP were the most prominent domestic initiatives designed to reduce GHG emissions proposed by the Clinton Administration in the first generation of US climate change policymaking. Other proposals were made including the US$6.3 climate change technology initiative (CCTI) in 1998. It was comprised of a variety of tax incentives and R&D expenditures. The CCTI was designed to improve energy efficiency and stimulate the development and deployment of lower and non-emitting fuels. The Republican Congress, which came to power in the mid-term elections of 1994, was never very supportive of these proposals.

International Climate Policy

My intent for this section is to provide a general review of the negotiations, events, and decisions that resulted in the KP, a treaty that was never presented to the US Senate for ratification. Others who know far more about the process than me have addressed this topic countless times. However, it is important to review some of the seminal decisions that were reached during this period in the international negotiations because it resulted in the KP and the carbon markets that dominated international climate change policy 1.0. An assessment of these decisions

---

[26] Clinton, President W.J., Vice President A. Gore Jr. *Climate Change Action Plan.* PP. 26–27 and A II-1 to A II-4.

is critical to understanding the reasons for the demise of the KP and to provide the foundation for a more effective and enduring successor agreement.

Key Milestones and Decisions in the Negotiation Process

This section makes several references to the UNFCCC of 1992, known by some as the Rio Treaty and otherwise referred to as 'the convention'. It was ratified by the US Senate in 1992 and entered into force on 21 March 1994. In short, the UNFCCC represented the global community's first effort to develop an international framework to address climate change. It is the umbrella under which the KP was negotiated. Among other things, it included an objective for 'stabilization of GHG concentrations in the atmosphere at a level that would prevent dangerous anthropogenic inter-ference with the climate system',[27] a goal for Annex I Parties to return their GHG emissions to 1990 levels individually or jointly,[28] the concept of JI,[29] and the roles and responsibilities of developed and developing countries in the effort to address climate change.

The convention also established that developed countries would take the lead in reducing their GHG emissions because of their historical and current emissions, that developing countries per capita GHG emissions were low, and that developing countries would necessarily increase emis-sions to support their continued development. As such the convention stated, 'The Parties should protect the climate system for the benefit of present and future generations of humankind, on the basis of equity and in accordance with their common but differentiated responsibilities and respective capabilities. Accordingly, the developed country Parties should take the lead in combating climate change and the adverse effects thereof'.[30]

---

[27] *UNFCCC.* Article 2.
[28] *UNFCCC.* Article 4 2. (b).
[29] *UNFCCC.* Article 4 2. (a).
[30] *UNFCCC.* Article 3.

It also established the COP to oversee the implementation of the Rio Treaty and any other related agreements that it may adopt.[31] The COP, which met annually beginning in 1995, was the body tasked with fleshing out the key details that were only vaguely alluded to in the UNFCCC. Decisions taken at these negotiations resulted in the KP. A general review of the key decisions follows.

COP-1: 28 March–7 April 1995

The road to the KP began in earnest at COP-1, which was held in Berlin, Germany in 1995. The UNFCCC included language for COP-1 to review the adequacy of commitments included in Article 4.2 (a) and (b) regarding Annex I Parties' policies and measures, and mitigation including the return of GHG emissions to 1990 levels, among others.[32] Based on the review, the COP 'shall take appropriate action' that could result in an amendment of Annex I Parties' commitments.[33] It also called for COP-1 to develop criteria for JI.[34]

COP-1 took several important decisions which were included in what became known as the Berlin Mandate. Based on the review cited above, it decided that Annex I Parties' progress made in achieving their commitments had been inadequate.[35] The Parties agreed to establish a process to strengthen Annex I Parties' commitments in the period after 2000 by adopting a protocol or other legal instrument.[36] The process would aim for Annex I Parties to 'elaborate policies and measures' and 'to set quantified limitation and reduction objectives within specified time frames, such as 2005, 2010 and 2020, for their anthropogenic emissions by sources and removals by sinks of GHGs not controlled by the Montreal Protocol'.[37] The decision also indicated that work should be completed

---

[31] *UNFCCC.* Article 7.

[32] *UNFCCC.* Article 4 2. (d).

[33] *UNFCCC.* Article 4 (2) (d).

[34] *UNFCCC.* Article 4 (2) (d).

[35] The First Session. *Report Of The Conference Of The Parties On Its First Session, Held At Berlin From 28 March To 7 April 1995, The Berlin Mandate.* Decision 1/CP.1. United Nations. 1995

[36] *The Berlin Mandate.* Decision 1/CP.1. United Nations. 1995.

[37] *The Berlin Mandate.* Decision 1/CP.1. 1995.

early in 1997 with the goal of adopting the results at COP-3.[38] This was the first formal decision on the road to developed countries agreeing to binding GHG reduction targets and timetables. This was a big deal given that the UNFCCC's targets had previously been voluntary in nature.

As important as the decision for Annex I Parties to set quantified limitation and reduction objectives for GHG emissions, the Berlin Mandate made it clear that developing countries would not be expected to make any new commitments from those included in the UNFCCC.[39] These decisions, made prior to 1992, altered the course of international climate diplomacy for the better part of 20 years.

Another important decision taken in the Berlin Mandate was to develop a pilot initiative called AIJ.[40] This was designed to create experience with JI projects. The decision also prohibited Parties to the convention from receiving any credits for GHG emissions that were reduced or sequestered during AIJ's pilot phase.[41]

The US business community strongly disliked this package of decisions. One cannot overstate the disdain they held for the Berlin Mandate, and the influence this had in setting the tone for the debate in climate change 1.0 in the US. The decisions at COP-1 unified the conservative faction of industry that would oppose any action to address climate change and the moderates that were skeptical but willing to work with the Administration. Business was concerned with the Administration's support for binding targets and timetables for GHG emissions for developed countries and the lack of developing country commitments. The concern centered on the environmental and competiveness implications of exempting developing countries from undertaking any new commitments. Industry that supported the market mechanisms also believed that the prohibition from receiving any credits during the AIJ pilot phase would serve as a disincentive for their companies to invest in overseas emission reduction projects.

---

[38] *The Berlin Mandate.* Decision 1/CP.1. 1995.

[39] *The Berlin Mandate.* Decision 1/CP.1. 1995. Many of the developing country commitments are included in Article 4 1. of the UNFCCC.

[40] *Activities implemented jointly under the pilot phase.* Decision 5/CP.1. 1995.

[41] *Activities implemented jointly under the pilot phase.* Decision 5/CP.1. 1995.

COP-2: 8–19 July 1996

At COP-2 in Geneva, the Heads of the delegation instructed their representatives to accelerate negotiations on an agreement based on the Berlin Mandate that could be adopted at COP-3 in Kyoto.[42]

Prior to the meeting, Tim Wirth, the Under Secretary of State for Global Affairs and formerly a US Senator, communicated that the US would accept a legally binding agreement with emissions limits if others did.[43] The Under Secretary Wirth also communicated that the US would, 'continue to seek market based solutions that were flexible and cost-effective'.[44] Another pivotal tenet in the US position articulated by Wirth was to 'lay the foundation for continuing progress by all nations in the future…[because] all nations—developed and developing—must contribute to the solution to this challenge'.[45] The Administration recognized the political fallout that resulted from exempting developing countries from undertaking new commitments from what they had agreed to in the UNFCCC.

The US position on what it would accept in an international climate change agreement that was to be completed at COP-3 was becoming clearer. It would accept binding targets and timetables for GHG emissions, a market-based policy framework, and some degree of developing country participation that was not defined. Although not stated explicitly, the US viewed the negotiation on a continuum. The stringency of the target that the US would accept was contingent on the level of flexibility in the implementation of its commitment. In other words, the US would likely accept a more stringent target if the agreement provided significant flexibility in implementation and included a market-based policy framework.

---

[42] *Report of the Conference of the Parties On Its Second Session, Held At Geneva From 8 to 19 July 1996, The Geneva Ministerial Declaration.* United Nations. 1996.

[43] Under Secretary for Global Affairs Tim Wirth before the Second Conference of the Parties Framework Convention on Climate Change, Geneva Switzerland, 17 July 1996. This and the following two citations were sourced from Royden, A. (2010). 'U.S. Climate Change Policy Under President Clinton: A Look Back', 32 *Golden Gate U. L. rev.* 32(4): 468–477.

[44] Under Secretary for Global Affairs Tim Wirth.

[45] Under Secretary for Global Affairs Tim Wirth.

COP-3: 1–11 December 1997

In the run up to Kyoto, domestic politics would intervene in the international negotiations in a big way. On 25 July 1997, approximately four months prior to the beginning of COP-3 in Kyoto, the US Senate passed a non-binding resolution, by a 95–0 margin, which was known as Byrd Hagel, in recognition of its two co-sponsors, Senators Robert Byrd (D-WV) and Chuck Hagel (R-NE). The language of the resolution is well-known. It says, 'Resolved, that it is the sense of the Senate that –

(1) the United States should not be a signatory to any protocol to, or other agreement regarding, the United Nations Framework Convention on Climate Change of 1992, at negotiations in Kyoto in December 1997, or thereafter, which would—

(A) mandate new commitments to limit or reduce greenhouse gas emissions for the Annex I Parties, unless the protocol or other agreement also mandates new specific scheduled commitments to limit or reduce greenhouse gas emissions for Developing Country Parties within the same compliance period, or

(B) would result in serious harm to the economy of the United States; and

(2) any such protocol or other agreement which would require the advice and consent of the Senate to ratification should be accompanied by a detailed explanation of any legislation or regulatory actions that may be required to implement the protocol or other agreement and should also be accompanied by an analysis of the detailed financial costs and other impacts on the economy of the United States which would be incurred by the implementation of the protocol or other agreement'.[46]

---

[46] Senator R. C. Byrd et al. *S. Res. 98, Expressing the sense of the Senate regarding the conditions for the United States becoming a signatory to any international agreement on greenhouse gas emissions under the United Nations Framework Convention on Climate Change.* Available at Government Printing Office Web Site: http://www.gpo.gov/fdsys/pkg/BILLS-105sres98ats/pdf/BILLS-105sres98ats.pdf (2 December 2015).

In simple terms, the Byrd Hagel resolution was in total conflict to the principles that were driving the international negotiating process. The Senators were strongly opposed to the provisions in the Berlin Mandate and Geneva Ministerial Declaration that put the US on a path to agree to mandatory GHG emission reduction targets and timetables while exempting developing countries from similar commitments. It did not matter that the US Senate had ratified the UNFCCC in 1992, which established that Annex I countries would take the lead, albeit in a voluntary fashion, in reducing GHG emissions. The difference was that the GHG reduction obligations that would be agreed to by the US at COP-3 would be legal in nature, as opposed to the voluntary goals included in the UNFCCC.

Although many believed that the floor debate attempting to clarify the resolution's meaning provided US negotiators some flexibility, the US would be in a strait jacket at COP-3 no matter how it was spun.[47] The debate on the senate floor attempted to clarify that the resolution did not require developing countries to achieve the same level of emission reductions as the US, but they would be expected to make some type of commitment in Kyoto regarding their GHG emissions in the same time frame as the US. Regardless, it was highly unlikely that developing countries would make any new commitments than what they had previously agreed to. Because of this, the treaty was arguably dead in the US before its contents were even known.

The US submitted a draft protocol framework in January of 1997 in preparation for Kyoto. It included several measures requiring developing countries to take no regrets measures to mitigate their GHG emissions, to prepare annual inventories and report steps to reduce their emissions, and to establish a process for reviewing developing country reports and improving their emission reduction strategies. It also included a provision that would have required all Parties to adopt GHG emissions targets by a certain date and included an automatic mechanism that would have imposed GHG emissions obligation on all Parties based on agreed criteria.[48]

---

[47] Harris, P. G. (1999), 'Common But Differentiated Responsibility: The Kyoto Protocol And US Policy', 7 *NYU Envtl L.J.* 27.

[48] US Draft Protocol Framework. 1997.

The one remaining event in the run up to Kyoto was a speech by President Clinton, in October 1997, detailing the US positions for the upcoming negotiations in Kyoto. In the speech, he committed the US to return its GHG emissions in 2008–2012 to 1990 levels. He also supported joint implementation and emissions trading to meet these limits and said that the 'US would not assume binding obligations unless key developing countries meaningfully participate in this effort'.[49]

The KP was agreed to on the last night of COP-3 on 11 December 1997.[50] It is important to note that the Administration was successful in achieving many of its objectives in the negotiations, particularly regarding the initial elements necessary to create the carbon markets. However, there was no chance the US Senate would ratify the agreement.

From a political and substantive perspective, the most glaring weaknesses of the KP was the developing country exclusion from GHG targets and a series of related issues that many believed undermined the US competitiveness. In addition to the developing country issue, the US accepted a reduction target of 7 % below 1990 levels in 2008–2012, which was far more stringent than proposed by President Clinton. Many viewed this commitment as unrealistic and too expensive given that the US was in the midst of a period of rapid economic growth. This concern was magnified by the impact of other details that some believed would provide firms in the European Community with an advantage over the US in international markets. These included the view that its target was not as ambitious as the US target. And this was exacerbated by a provision that authorized the community to achieve its aggregate target by distributing the burden among its 15 pre-2004 members as it saw fit.[51] This resulted in a 1998 agreement called the burden sharing agreement (BSA) or bubble, as it was dubbed. This rankled many in the US who did not think that the European Community should be provided this level of flexibility,

---

[49] William J. Clinton: 'Remarks at the National Geographic Society,' October 22, 1997. Online by Gerhard Peters and John T. Woolley, *The American Presidency Project*. http://www.presidency.ucsb.edu/ws/?pid=53442.

[50] *Kyoto Protocol to the United Nations Framework Convention on Climate Change*. United Nations. 1998.

[51] *Kyoto Protocol*. Article 4.

particularly as it had opposed the US on such issues during the negotiation. Chapter 4 discusses these issues in greater detail.

Key Features of the KP

It is not necessary to describe the entire KP. However, because a major focus of the book is assessing the performance of the carbon markets, a brief description of the provisions in the agreement that created the market demand for GHG emissions reductions and those designed to provide the supply to meet demand follows.

The Creation of Market Demand

Article 3 of the KP created the market demand for GHG reductions by imposing binding reduction targets on developed countries. It stated that the GHG emissions of 39 developed countries included in Annex I of the convention should not exceed the limit included in Annex B of the KP from 2008 to 2012.[52] The limits, known as 'quantified emission limitation or reduction commitments' were expressed as a percentage of a country's 1990 baseline. The US limit was 93%, which required it to reduce GHG emissions 7% below 1990 levels.[53] The EU agreed to a target of 92% of its 1990 baseline.[54]

On an aggregate basis, the 37 countries GHG emissions that originally ratified the KP would be approximately 5% below 1990 levels in 2008–2012, which became known as the first commitment period. The US never ratified the KP and Australia did not ratify it prior to it taking effect in February of 2005. Although Canada did ratify, it did not take any actions to achieve its targets prior to withdrawing from the KP in 2011. Natsource estimated that these targets created demand for approximately 3.8 billion tons of GHG emission reductions in Europe, Canada, and Japan during the 2008–2012 commitment period.

---

[52] *Kyoto Protocol.* Article 3.

[53] *Kyoto Protocol.* Annex B.

[54] *Kyoto Protocol.* Annex B.

Project-Based Mechanisms to Create Supply

Articles 6 and 12 of the KP created two project-based mechanisms: JI[55] and the CDM,[56] respectively.

The CDM

The CDM was a hybrid that resulted from a concept developed by Brazil, which would have required developed countries to pay into a fund for non-compliance for the benefit of developing countries and JI. Its objectives were to contribute to the host country's sustainable development and to assist Annex I Parties (those that agreed to emissions limits) comply with their GHG emissions targets.[57] An Executive Board (EB) would supervise the CDM.[58] Projects would be certified by independent entities called operational entities, which came to be known as Designated Operational Entities or DOEs.[59] To create carbon credits, known as certified emission reductions (CERs), CDM projects were required to achieve reductions 'that are additional to any that would occur in the absence of the project activity'.[60] This concept, known as additionality, has contributed to offset systems inefficiency, higher transaction costs, and increased controversy. To demonstrate additionality, project developers are required to develop a counterfactual scenario of what GHG emissions would be in a business as usual (BAU) scenario in the absence of the project activity. They then need to demonstrate that emission reductions were achieved from BAU estimates by the project activity. The process of developing the BAU GHG emissions scenario and proving that the project reduced emissions from that level is always contentious. Projects would be independently audited and verified.[61]

---

[55] *Kyoto Protocol.* Article 6.
[56] *Kyoto Protocol.* Article 12.
[57] *Kyoto Protocol.* Article 12 (2).
[58] *Kyoto Protocol.* Article 12 (4).
[59] *Kyoto Protocol.* Article 12 (5).
[60] Kyoto Protocol, Article 12 (5) (c).
[61] *Kyoto Protocol.* Article 12 (7).

And in an attempt to stimulate early action, 2000–2007 CERs could be used for compliance.[62]

Simply put, CERS created by CDM projects could be used by Annex I countries to comply with their Kyoto targets. It became an important component of the global carbon markets established by the KP and is the focus of significant analysis.

JI

JI was changed from its original construct. Instead of authorizing Annex I Parties and investors to undertake emission reduction projects in developing countries, it authorized such projects to be implemented in other Annex I countries. It was thought that JI projects would be undertaken primarily in countries located in Central and Eastern Europe with economies in transition (EIT) such as the Czech Republic, Hungary, Poland, Russia, and Ukraine. The emission reduction units (ERUs) that would be created by such projects could be used by Annex I Parties for compliance with their KP targets. To maintain the integrity of the Annex I emissions cap, the ERUs, which would be added to the acquiring Parties' national allotment, called an assigned amount, and would be subtracted from the transferring Parties' allotment. This was an important distinction between CDM and JI.

Article 17 also authorized international emissions trading among Annex I Parties.[63] Chapter 3 details early market activity and the rules that were developed in the international negotiations to implement the CDM.

In the initial construct of the carbon markets, many envisioned that the US, the Annex I Party with the largest reduction requirement, would purchase a portion of Russia's and the Ukraine's allocation within the construct of the international emissions trading provisions in the KP. They would have a large surplus to sell because their targets were set at 1990 levels although their GHG emissions declined by more than

---

[62] *Kyoto Protocol.* Article 12 (10).
[63] *Kyoto Protocol.* Article 17.

50 % from the 1990 base year to 2000.[64] Many were concerned that this compromised the integrity of the system because these countries would be rewarded for selling the surplus caused by the fall of the former Soviet Union without making any effort to reduce its GHG emissions. The generous allocation was quickly labeled 'hot air'. To market enthusiasts, it was necessary to provide liquidity, jumpstart the market, and manage US compliance costs.

International emissions trading, and the reductions created by JI and CDM projects (ERUs and CERs, respectively) would be used to supplement Annex I Parties' domestic policies and measures to meet their GHG reduction targets. The US government analysis of the KP showed that these mechanisms would significantly reduce the nation's cost to comply with its emission reduction requirements.[65] These cost savings were enabled by the opportunity to make large-scale reductions in developing nations and industrialized countries with EITs that were much cheaper to achieve than in the US and other more efficient nations.

The US Reacts to the KP

Following agreement of the KP, the Administration continued to aggressively negotiate the details necessary for its implementation and to create the carbon market. However, these negotiations focused primarily on technical details and the long-standing disputes between the US and Europe on such issues as carbon sinks, the use of markets, and flexibility, and between the US and developing countries on their role in the international regime. In fact, COP-6, held in 2000, collapsed over substantial disagreement over the role of carbon sinks.[66] Little progress was made on the issue of developing country commitments.

---

[64] This information on the Russian Federation's and Ukraine's GHG emissions are included in the GHG emission profiles for Annex I Parties and major groups in the GHG data section of the UNFCCC website. Available on the UNFCCC website at: http://unfccc.int/ghg_data/ghg_data_unfccc/items/4146.php.

[65] Administrations Economic Analysis. *The Kyoto Protocol and the President's Policies to Address Climate Change*. 1998.

[66] Royden, A. (2010). 'U.S. Climate Change Policy Under President Clinton: A Look Back', 32 *Golden Gate U. L. rev.* 32(4): 468–477.

At home, strong disagreements continued to stymie efforts to develop a domestic response to climate change and over Kyoto. The Administration continued to attempt to stimulate the development and deployment of low and non-emitting technologies, and Republicans began to develop alternatives to the KP that consisted primarily of voluntary GHG reduction efforts and technology initiatives. Major disagreements also emerged between the Administration and its opponents over the costs of implementing Kyoto. The Administration estimated Kyoto would raise annual household energy bills cost in 10 years between US$70 and US$110.[67] In contrast, the Energy Information Administration (EIA), an independent analytical arm of the DOE that created the respected annual energy outlooks (AEO), estimated Kyoto's impacts would be far greater. It concluded that the agreement would result in delivered energy costs 17–83% higher than 2010 projections, increase gasoline costs between 11 and 53% and cause the loss of 10,000–43,000 coal-miners jobs.[68,69] It should be noted that the EIA analysis did not consider the potential cost-saving benefits of the flexible mechanisms because of the lack of detail regarding their implementation at the time the analysis was undertaken. And Kyoto opponents were furious when the Administration signed the KP at COP-4 in 1998. They viewed this action as a blatant disregard of the Byrd Hagel resolution. All this played out at the same time as President Clinton's impeachment; an event that effectively curtailed his ability to engage consistently in the debate.

There can be no denying that the Clinton Administration did its best in negotiating the KP. Without the US, there never would have been the carbon markets that many countries utilized in an attempt to implement their commitments. However, the Administration could never secure the meaningful participation of key developing countries necessary to move it forward. The path to this stalemate started with the agreement of the UNFCCC and was exacerbated by the Berlin Mandate. The Administration continued to call for meaningful developing country

---

[67] Administrations Economic Analysis. *The Kyoto Protocol and the President's Policies.* P. iv.
[68] US Energy Information Administration. *What does the Kyoto Protocol Mean to U.S. Energy Markets and the U.S. Economy?* Report number: SR/OIAF/98-03. (Department of Energy) 1998.
[69] See pages 451–453 in Royden.

participation, but the prior decisions going back to 1992 were not going to be overcome. Regardless of what many think, and some of my former colleagues in the Clinton Administration may disagree, but there was never a path to senate ratification because of the decisions regarding developing countries that were made in 1992 and 1995.

The adoption of the KP was both a high and low point in the Clinton Administration's efforts to create an enduring global framework to address climate change. It was the first international agreement designed to address climate change. Nearly 40 countries accepted binding GHG emission reduction targets for the first time. The framework of a global market was created and many believed it would unleash the creativity and innovation necessary to mobilize the capital required to address the century scale challenge of climate change. On the flip side, it was clear that Kyoto in its original form was dead on arrival in the US.

The Market Reacts to the KP

Although some were convinced that a market would be created based on the JI concept as early as 1990, interest increased significantly in the aftermath of the Berlin Mandate and the inexorable march toward Kyoto, which was dominated by discussions of market mechanisms in the US and internationally. Many were convinced that an international agreement would be completed that included binding targets for GHG emissions and some type of project-based market mechanism modeled after pilot initiatives such as USIJI, Canada's Pilot Emission Reduction Trading Program (PERT), AIJ, and others.

Although it was unlikely that the US would ratify the KP, that its entry into force was uncertain, and that much work needed to be done to make market mechanisms a reality, companies began to organize to participate in what many believed would become a trillion dollar commodity market. Natsource LLC was a mid-sized introductory broker in emissions and energy markets when the KP was completed. Jack Cogen, who had previously headed Eurobrokers natural gas brokerage, founded Natsource in 1994. He was a recognized expert in emissions and energy markets and helped create the rules for natural gas derivatives trading.

As a broker, Natsource introduced counter-parties to transact emissions permits/allowances for sulfur dioxide (SO_2) in the US acid rain trading program and other environmental commodities such as oxides of nitrogen (NOx) to address other air quality problems such as ozone formation. It also arranged some of the first transactions for renewable energy credits (RECs) and brokered natural gas, power, and coal. Because of this, the company had knowledge of environmental and energy markets and regulation, and it had a long list of customers, particularly in the power and industrial sectors that would be affected by any climate deal. The company also had Japanese investors. During this period, delegations of Japanese companies began a dialogue with Natsource to increase their understanding of environmental markets.

In recognition of customer interest in climate policy, and the potential scale of the market, Natsource began to prepare for the future by hiring carbon brokers following agreement of the KP. Their job was to build relationships with potential market participants, and begin to create the company's presence internationally by attending the many conferences that were being held to discuss climate change and carbon markets. Other companies and entrepreneurs also began to explore the potential to create businesses to participate in the undefined market. However, in December 1997, with the KPs prospects uncertain, and shape of the market undefined, there was no understanding of what a GHG business might look like.

Some were of the view that the markets would flourish once the KP was agreed to. Coming from the world of public policy, I was naïve in this regard. Large-scale activity in the carbon market would be dependent on progress on the commercial and political fronts. This section describes some of the early activities in the project-based market and initiatives that helped move it forward.

Early Activity and Evolution of the Carbon Market

As in any market, sellers and buyers are required. Some project developers began to attempt to sell their GHG emission reductions while a few prospective buyers were interested in learning by doing. But this was at a time in which there was no clarity governing the creation of CERs and ERUs, or any emission reduction created by a project and their use following agreement of the KP. There was an understanding that a project would need to create emission reductions that were 'additional' to what would have occurred in the absence of the project. But this was not understood in practice. Negotiators continued in their attempt to define additionality and to develop rules to govern CDM and JI.

In the absence of formal rules, the pilot initiatives and market activity began to spur the development of a common set of quality criteria to guide the development of GHG reduction projects.[70] Developers used the criteria that were included in the pilot programs as guides to implementing their projects. To address developers' interest in selling reductions and buyers' need to eventually use the market as a compliance option, Natsource and other brokers began to incorporate criteria in term sheets marketing project-based reductions.[71] These early activities were important in the initial stage of the market for project-based reductions. However, there were far more conversations than transactions. The risks inherent in a transaction for GHG emission reductions of any scale far outweighed the benefits for buyers. In short, they would be spending money for a commodity that could have no future value.

There were significant risks involved in GHG emission reduction projects for sellers and buyers (and there continues to be). Sellers needed to develop GHG emission reduction projects that would conform to regulatory and environmental criteria both in the host country and at the international level, and they had to develop the necessary project documentation. They needed to understand how to apply technologies and learn how to operate their projects. And sellers needed to develop a set of commercial terms that could be incorporated in contracts to allocate project-related risks and to secure prices for GHG emissions reductions that would assure a project's economic viability. Buyers also confronted a set of risks when contracting for project-based reductions. First and foremost, they needed to assess the risks that a project would

---

[70] Rosenzweig, R., M. Varilek, B. Feldman R. Kuppalli, and J. Janssen. *The Emerging International Greenhouse Gas Market.* Formerly Pew Center on Global Climate Change and Currently the Center for Climate and Energy Solutions. 2002. PP. 4–5.

[71] For an example of a sample term sheet, see PP. 53–55 in R. Rosenzweig above in previous note. Available at the Center for Climate and Energy Solutions Website at: http://www.c2es.org/.

deliver contracted volumes given that they would be used to comply with a GHG reduction obligation. The risk assessment needed to be multifaceted and consider: the host country's economic climate, credit-worthiness of the seller, regulations at the host country and international level, technology, and the project's ability to operate. Similar to any other commercial endeavor, these risks needed to be allocated between buyers and sellers in a purchase and sales agreement that came to be known as an emission reduction purchase agreement (ERPA).

To mitigate these risks, participants required far more clarity regarding the rules of the road that would govern the implementation of the mechanisms. How would this occur? Would it result from progress in the negotiation of the rules or from investment in GHG reduction projects? As we will see, it took both.

Confidence in the mechanisms began to increase as a result of invest-ments by the World Bank (WB) and several governments in what were then called 'candidate' CDM and JI projects. They disseminated the knowledge and lessons learned to negotiators and others to inform the development of rules necessary to implement the mechanisms. The two most prominent initiatives during this period were the WB Prototype Carbon Fund's (PCF) and the Netherlands' purchases of project-based reductions.

The WBPCF

The PCF was established in 2000. It raised US$180 million from six gov-ernments and 16 private firms to invest in CDM and JI projects. The PCF had several goals that included illustrating how the project-based mecha-nisms could contribute to sustainable development in the host country and lower the costs of Annex I Parties compliance; providing learning to Parties, private firms, and others to achieve emission reductions by using CDM and JI; and, demonstrating how the WB could mobilize resources for borrowing countries and address environmental problems through market mechanisms.[72] Another goal was to communicate the knowledge

---

[72] The World Bank. *Prototype Carbon Fund.* Available from: http://www.worldbank.org/en/topic/climatechange/brief/world-bank-carbon-funds-facilities. [Accessed 3 December 2015].

learned from the development of GHG emission reduction projects and contracting for their reductions to negotiators developing the rules to govern the mechanisms.[73] The PCF ultimately created a portfolio of 24 CDM and JI projects.[74]

Significant concern was expressed during this period regarding the WB's appropriate role in the carbon market and whether it would crowd out the private sector. Although somewhat concerned given the WB's influence and resources, Natsource decided that it was in its best interest to develop a working relationship with the bank, rather than bashing it, which our competitors frequently did. At Natsource, the Advisory and Research business that I headed worked with the PCF's research arm to invent an annual publication called 'State and Trends of the Carbon Market' beginning in 2001 and which we continued to work on through 2008.[75] These were annual publications that continue to be published today. They provided data on such market issues as volumes transacted and their dollar amount, prices paid for project-based reductions, the location of buyers and sellers, the types of technologies that created the emission reductions, contract types, and other market developments. More importantly, Natsource was the largest buyer in the Umbrella Carbon Facility (UCF), a WB syndicate that was the world's first and potentially only US$1 billion CDM deal ever completed. This transaction is the subject of discussion in Chap. 3.

The WB played a constructive role in the early days of the market. The lack of rules governing CDM and JI was a significant disincentive to investment. The private sector would not deploy large amounts of capital until there was greater certainty. The PCF's learning-by-doing approach to investing in GHG emission reduction projects helped increase confidence in the mechanisms and provided negotiators with useful information that led to progress in crafting the rules to govern

---

[73] LeCocq, F. 'Pioneering Transactions, Catalyzing Markets, And Building Capacity: The Prototype Carbon Fund Contributions to Climate Policies', *Amer. J. Agr. Econ.* 2003 85 (3) August 2003: 703–707.

[74] The World Bank. *Prototype Carbon Fund.*

[75] Rosenzweig, R., D. Forrister. Natsource Compiles First Comprehensive Analysis of the Greenhouse Gas Trading Market. [Press Release] 6 August 2001. I have been unable to locate the first State and Trends of the Carbon Markets completed by the PCF in 2001.

them. The WB also developed funds to attempt to prove out sequestration[76] and facilitate investment in carbon reduction projects in poor communities (known as the Community Development Carbon Fund).[77] The bank was attempting to build confidence in market segments that were too risky for the private sector to invest in at the time. The WB continues in this role today. It has launched several new initiatives to build confidence in market mechanisms in the current era of policy-making.[78]

Many, including myself believed the WB overstepped its role in the initial era of the carbon market. It established four country funds and one European fund to purchase CERs and ERUs that the government could use for compliance.[79] These activities crowded out the private sector. The bank had several advantages over entrepreneurial firms like Natsource and others. It had enormous resources at its disposal, could offer several benefits to host countries that the private sector could not, and had long relationships with host country governments. It used these advantages to secure business.

Government Initiatives

The government of the Netherlands was also an early carbon market participant. It developed two procurements: the Certified Emissions Reduction Procurement Tender (CERUPT) and the Emission Reduction Unit Procurement Tender (ERUPT), to purchase CERs and ERUs that it would use to comply with its KP emission reduction obligations. The Netherlands' goals were to utilize the market to purchase reductions to

---

[76] The World Bank. *BioCarbon Fund*. Available from: http://www.worldbank.org/en/topic/climatechange/brief/world-bank-carbon-funds-facilities [Accessed 3 December 2015].

[77] The World Bank. *Community Development Carbon Fund*. Available from: http://www.worldbank.org/en/topic/climatechange/brief/world-bank-carbon-funds-facilities [Accessed 3 December 2015].

[78] The World Bank. *World Bank Carbon Funds and Facilities*. Available from: http://www.worldbank.org/en/topic/climatechange/brief/world-bank-carbon-funds-facilities [Accessed 3 December 2015].

[79] The World Bank. *World Bank Carbon Funds and Facilities*. Available from: http://www.worldbank.org/en/topic/climatechange/brief/world-bank-carbon-funds-facilities [Accessed 3 December 2015].

comply with 50% of its GHG emissions reduction obligation, and assist in the market's development.[80] The Netherlands also hired other entities including the WB to serve as a purchasing agent for CERS and ERUs.

Although on a smaller scale, several other European governments implemented similar initiatives to purchase CERs and ERUs or outsourced a portion of their Kyoto obligations.[81] Later in the market's evolution, the government of Portugal participated in a Natsource fund, the Natsource Carbon Asset Pool (NATCAP). Given the ideological concerns with the appropriate role of government in the US, it is difficult to imagine the US participating in the carbon market similar to the way European governments did. Such initiatives appeared logical. Government's purchases of CERs and ERUs spread the cost of KP compliance and risk of investing in the mechanisms throughout the society.

Regardless of the cost of these activities and results they ultimately achieved, the PCF and the early efforts European governments helped create the conditions for the markets to move forward.

---

[80] Henkemens, M. *Dutch lessons as GHG buyer.* [Lecture] New York. 25 June 2004.

[81] de Dominicis, A. *Carbon investment funds: growing faster.* Caisse des Depots. Research Report No. 7. 2005 November.

3

The US Says No While the Carbon Market Moves Forward

The US Retrenches

The eight years of the Bush Administration were a study in contrasts in climate change policy at home and internationally. In a speech on energy policy that took place in Saginaw, Michigan, on 29 September 2000, Candidate Bush said, 'We will require all power plants to meet clean air standards in order to reduce emissions of sulfur dioxide, nitrogen oxide, mercury and carbon dioxide within a reasonable period of time'.[1] Most interpreted this as a commitment to address climate change.

Four Republican Senators sought to clarify the policy in a letter to President Bush dated 6 March 2001.[2] President George W. Bush's response, dated 13 March 2001, dramatically altered the climate change policy landscape at home and in the international negotiations. He wrote, 'As you know I oppose the Kyoto Protocol because it exempts 80

---

[1] S. Borenstein. *Bush Changes Pledges on Emissions*. Philadelphia Inquirer. 2001 March 14.
[2] Hagel, C., L. Craig, J. Helms, P. Roberts. Letter to President Bush seeking clarification on the Administrations climate change policy. 6 March 2001.

© The Author(s) 2016
R.H. Rosenzweig, *Global Climate Change Policy and Carbon Markets*, Energy, Climate and the Environment,
DOI 10.1057/978-1-137-56051-3_3

per cent of the world, including major population centers such as China and India, from compliance, and would cause serious harm to the US economy. The US Senate's vote, 95–0, shows that there is a clear consensus, that the Kyoto Protocol is an unfair and ineffective means of addressing global climate change concerns.'[3] In the same letter, President Bush expressed his opposition to requiring mandatory reductions of power plant CO_2 emissions,[4] reversing the pledge he made a little more than five months ago during the presidential campaign.

President Bush's decision to oppose the KP has been written about many times. I will not provide my views on who and what drove the decision. However, its ramifications were felt at home and abroad. Domestically, development of climate policy at the federal level came to a halt. The Administration never put forward any serious initiatives to reduce national GHG emissions during its eight years in power. Legislative proposals to reduce pollutants from power plants, including CO_2, and to create a GHG cap-and-trade system were introduced and received some consideration during President Bush's two terms, but there was virtually no chance they would become law. Partially due to frustration with the lack of action at the federal level, some states developed their own responses to climate change beginning in 2005, initiating bottom-up climate policies in the US.

At the international level, the Bush Administration's decision dramatically reduced the potential for the KP to enter into force because of the requirement that 55 Parties to the Convention, accounting for 55% of 1990 Annex I CO_2 emissions, ratify the agreement.[5] Because the US accounted for 34%[6] of 1990 Annex I CO_2 emissions, achieving this threshold would be extremely difficult. Nearly all other 1990 Annex

---

[3] G. W. Bush: 'Letter to Members of the Senate on the Kyoto Protocol on Climate Change,' March 13, 2001. Online by Gerhard Peters and John T. Woolley, *The American Presidency Project.* http://www.presidency.ucsb.edu/ws/?pid=45811

[4] G. W. Bush: 'Letter to Members of the Senate'.

[5] *Kyoto Protocol to the United Nations Framework Convention on Climate Change.* Article 25. United Nations. 1998.

[6] US CO_2 emissions in 1990 were 5.1 Gt CO_2 of the Annex I total of nearly 15.2 Gt CO_2 representing approximately 34%. This data was derived from the GHG emission profiles for Annex I Parties and major groups in the GHG data section of the UNFCCC website. Available on the UNFCCC website at: http://unfccc.int/ghg_data/ghg_data_unfccc/ghg_profiles/items/4625.php

I Parties would be required to ratify the KP for it to enter into force. Russia's emissions, which accounted for over 15% of 1990 Annex I emissions, had far less economic incentive to ratify with the largest buyer out of the market.[7] The Protocol could not enter into force if Russia did not ratify. There was also significant uncertainty as to whether Canada would ratify given the interrelationships between the Canadian and US economies. And Australia was always a wild card given its economy's reliance on coal and energy-intensive industries like steel and aluminum.

Following the US decision not to ratify, the KP covered approximately 33% of global GHG emissions.[8] There was no way it could be effective in addressing climate change. It was clear that a successor agreement covering a much larger percentage of global GHG emissions would need to be negotiated should the KP ever enter into force.

Although US leverage may have been at its highest point, it pulled the plug on the KP in a heavy-handed fashion. The reasons for this rejection no longer matter. However, it is clear that conservative elements of the energy sector were communicating their strong opposition at the time. The Administration may have misjudged the impact the decision would have. Perhaps they did not think it would be a big deal. There would be a few days of noise and then it would be forgotten. This was not the case. The decision to renounce Kyoto created a firestorm. It is still unclear to me why the Administration did not take a more conciliatory approach. They could have expressed their opposition to the objectionable provisions to the KP, such as the stringency of the US emission reduction targets and/or the lack of developing country commitments, and state its intention to make new proposals that would make the KP more acceptable at home and effective in addressing climate change. Although the US would have been subjected to major criticism for taking this path, some in the international community may have preferred negotiating changes

---

[7] Russia's CO_2 emissions in 1990 were 2.5 Gt CO_2 of the Annex I total of nearly 15.2 Gt CO_2 representing over 15%. This data was derived from the GHG emission profiles for Annex I Parties and major groups in the GHG data section of the UNFCCC website. Available on the UNFCCC website at: http://unfccc.int/ghg_data/ghg_data_unfccc/ghg_profiles/items/4625.php

[8] This percentage was derived from reviewing several data sources including the websites of the US Environmental Protection Agency, the UNFCCC, and the CAIT Climate Data Explorer developed by the World Resources Institute. A description of the calculations and sources are provided in a discussion of the Kyoto Protocol in Chap. 4.

to the agreement that could have made it more palatable to the world's largest emitter than an agreement with no US participation. Instead, the US decision, and the way it was presented, galvanized the international community to action.

The International Market Emerges

In the aftermath of the US stated opposition to the KP, the Parties continued to negotiate the details necessary to operationalize the Protocol and its market mechanisms.

The Marrakesh Accords: The CDM Rules Become Clearer

After agreement of the KP in 1997, the pilot initiatives cited in Chap. 2 helped to increase interest in CDM and JI projects. Although these early efforts were providing important lessons and increasing confidence, greater regulatory certainty regarding the operation of the mechanisms was essential to mobilizing large-scale investment necessary to reduce GHG emissions. It was against this backdrop that COP-7 was held in Marrakesh, Morocco, in 2001. Although it focused on many issues, a major emphasis was in developing the rules to govern the mechanisms.

The emphasis on the CDM is in recognition of its importance in jump-starting the market, the role it played in the 2005–2012 period, and that Natsource contracted over 100 million CERs. The CDM's prominence resulted from a provision authorizing the use of 2000–2007 vintage CERs for compliance in the 2008–2012 Kyoto Period.[9] The subsequent launch of the EU ETS provided CERs with compliance value in Phase 1, which ran from 2005 to 2007.

---

[9] *Kyoto Protocol to the United Nations Framework Convention on Climate Change.* Article 12 10. United Nations. 1998.

There are multiple goals in providing the information that follows. They are to enable the reader to understand how the rules governing the project cycle affected the CDM's performance and to set the stage for an analysis of such in Chap. 4. In addition, the information illustrates the challenge in developing an efficient offset mechanism capable of achieving large-scale mitigation at the project level.

Clean Development Mechanism Modalities and Procedures

This section identifies some of the key elements of the agreements reached at COP-7, which became known as the Marrakesh Accords. These include the requirements for Parties to participate in the mechanism, the roles and responsibilities of entities in administering and supervising it, and the important steps in the project cycle governing the creation and issuance of CERs.[10]

Eligibility Requirements

Parties were required to comply with a common set of 'eligibility requirements' to participate in CDM, JI, and international emissions trading. They included that (i) a Party calculate its assigned amount in accordance with prior COP decisions; (ii) a system be in place to estimate emissions by sources and removals by sinks of GHG emissions not controlled by the Montreal Protocol; (iii) a national registry be in place; and (iv) Parties have submitted its most recent inventory in compliance with COP requirements.[11]

---

[10] *Report Of The Conference Of The Parties On Its Seventh Session, Held at Marrakesh From 29 October to 10 November 2001, The Marrakesch Accords. Addendum, Part Two: Action Taken by the Conference of the Parties, Modalities and procedure for a clean development mechanism as defined in Article 12 of the Kyoto Protocol.* Decision 17/CP.7. FCCC/CP/2001/13/Add.2. United Nations. 2002.

[11] *Marrakesch Accords. Modalities and procedure for a clean development mechanism as defined in Article 12 of the Kyoto Protocol. Draft Decision -/CMP.1. Annex F. Participation Requirements.* FCCC/CP/2001/13/Add.2. United Nations. 2002.

Governance

The COP, serving as the Meeting of the Parties to the Kyoto Protocol (COP/MOP), would oversee the CDM.[12] It would be supervised by an Executive Board, or EB, as it was known.[13] The EB's administrative authorities and role in overseeing the project cycle are described throughout the COP decision and accompanying Annex detailing the rules that were agreed to.[14] Operational Entities, accredited by the EB, and which were known as Designated Operational Entities would undertake the technical work required to validate CDM project activities and to verify and certify that the GHG reductions were additional to what would have happened in the absence of the project activity.[15]

A Designated National Authority (DNA) from the project participants (PPs) and the host country were required to certify voluntary participation in the project and issue letters of approval (LOAs). Host country DNAs were also responsible for certifying that a project activity conformed to its sustainable development criteria.

Project Cycle

CDM project activities were required to conform to detailed guidance necessary to demonstrate additionality to earn CERs in the project cycle. Key steps in the process follow.

1. Validation and registration. This was the first part of the process.[16] It required PPs to develop a project design document (PDD) which would be independently reviewed and undergo validation by a designated operational entity. Among the most important information to be included in the PDD were a description of (i) the project; (ii) an

---

[12] *Marrakesch Accords. Annex B. Role of the Conference of the Parties Serving as the Meeting of the Parties.* United Nations. 2002.

[13] *Marrakesch Accords. Annex C. Executive Board.* United Nations. 2002.

[14] *Marrakesch Accords. Sections C, D. G. I. and J. of the Annex.* United Nations. 2002.

[15] *Marrakesch Accords. Annex E. Designated operational entities.* United Nations. 2002.

[16] *Marrakesch Accords. Annex G. Validation and registration.* United Nations. 2002.

approved methodology or a new methodology if there was not an applicable approved methodology in place that would be used; (iii) how the project activity would reduce GHG emissions below those that would have occurred under business as usual; (iv) a definition of BAU and how the project went beyond it; (v) what the project was doing differently from what was common in the industry; (vi) calculations to quantify the GHG reductions that were achieved; (vii) a monitoring plan to monitor project performance; and (viii) the environmental impacts of the project.[17]

The designated operational entity would develop a validation report and formally request the EB to register the project based on meeting the requirements above. Registration of a project was the formal acceptance by the EB of the validated project as a CDM project activity. Registration would be granted unless a Party in the project or three members of the EB requested a review.[18]

2. Verification and certification. Following registration, this was the next step in the project cycle. In general this required the designated operational entity to verify, on an ex post basis, that the PPs implemented the project and monitoring plan as stated in the PDD. Based on the review, it would verify the monitored volume of reductions achieved by the project activity during the verification period. Following verification, the designated operational entity would provide a verification report to the participants, the Parties, and the EB. It would then certify in a certification report to the same entities that the project activity achieved a verified amount of GHG emission reductions.[19]

3. Issuance. This was the final step in the project cycle. The certification report would constitute a formal request to the EB for issuance of CERs. The issuance would occur in 15 days unless one of Parties involved in the project or three members of the EB requested a review.[20]

---

[17] *Marrakesch Accords. Appendix B. Project design document.* United Nations. 2002.

[18] *Marrakesch Accords. Annex G. Validation and registration.* United Nations. 2002.

[19] *Marrakesch Accords. Annex I. Verification and certification.* United Nations. 2002.

[20] *Marrakesch Accords. Annex J. Issuance of certified emission reductions.* United Nations. 2002.

The Next Phase in Market Evolution

The carbon market evolved in different stages in response to political events and pilot offset initiatives. The first stages focused on the supply side of the market and began after the Berlin Mandate in 1995 and agreement of the KP in 1997. The WB, Netherlands, and other European government purchasing initiatives, which began around 2000, and agreements reached in Marrakesh in 2001 continued to establish the supply side of the market.

The missing piece was demand for GHG emission reductions. Market demand began to grow following the ratification of the KP by Japan and Canada in 2002. Both were slated to be large buyers. Analysis estimated that these countries would both need to reduce emissions by more than 30 % from 2010 BAU estimates to comply with their KP requirements.[21] If Russia ratified the KP, its condition for entry into force would be met.

In addition, the EU ETS was also being developed as the key component of EU's strategy to comply with its obligations under Article 3 of the KP. The EU ETS was expected to provide another source of demand for project developers. These dynamics created the conditions for increased CDM project development and for buyers to begin exploring the market in earnest.

The Evolution and Brief Description of the EU ETS

It is ironic that the EU ETS was the world's first and largest GHG cap-and-trade system developed to date given the sometimes bitter disputes between the EU and US in the international negotiations on the use of market mechanisms and flexibility to achieve climate policy objectives. On the other hand, it was entirely logical. The EU had previously attempted to implement a carbon tax in the 1990s. The tax was withdrawn in 1997 following strong industry opposition. Because of this, taxes did not appear to be an option for the EU to achieve its climate

---

[21] MacCracken, C. N., J. Edmonds, S. Kim and R. Sands, (1999), 'The Economics of the Kyoto Protocol", *The Cost of the Kyoto Protocol: A Multi-Model Evaluation, A Special issue of the Energy Journal*', 40. P. 40.

policy objectives. An ETS was the remaining market-based instrument to achieve large-scale reductions in GHG emissions.

Prior to a discussion of President Clinton's proposed BTU tax in Chap. 2, I provided a brief description of the theory behind taxes as a preferred instrument to achieve environmental objectives, including reductions in GHG emissions. Before turning to the evolution of the ETS and a review of its performance in 2005–2007, a similar description of some of the elements of a cap-and-trade program is presented.

The Basics of Cap-and-Trade

In general, cap-and-trade programs impose a fixed limit, or a cap, on a firm's emissions and generally provides them with the flexibility to determine how best to comply with the individual limits. In contrast to a tax, which provides certainty with respect to compliance costs, cap-and-trade provides certainty in achieving a fixed level of emissions. The firms covered by the program are generally required to surrender an amount of permits/allowances (usually equivalent to a ton of pollution) to regulators to cover their annual emissions or another proscribed period of time. In the US acid rain program, firms could comply with their limits by reducing their power plant's SO_2 emissions; by switching to natural gas or lower sulfur coal; installing flue gas desulfurization technology, known as scrubbers; or, through buying excess permits/allowances from other firms in the market. The costs saving benefits of the SO_2 cap-and-trade program were derived from a firm's ability to utilize the lowest cost compliance options. This meant that a firm could purchase permits/allowances from other firms at costs that were lower than cutting emissions in their own power plants. The opportunity for trade also provided an incentive for firms to continue to cut their emissions below their limits and sell them in the market.

In a GHG cap-and-trade program, the program can be limited to trade between covered sources, a closed system. Or, as in the case with the EU ETS during its first three phases, it can be an open system that allows covered sources to purchase offsets or emission reductions created by entities outside of the cap-and-trade system. Although the US acid rain and ozone programs included in the CAAA have been substantially modified since their inception, they achieved significant results, particularly in the early years of the programs.[22,23] Cap-and-trade systems also usually include a penalty for non-compliance. These are typically financial penalties that require a payment per ton for non-compliance and/or require non-compliant firms to pay back the amount of tons they were short with an interest payment.

Creation of the EU ETS

Following agreement on the KP, a series of papers were developed that led to the establishment of the EU ETS. The first indicated the potential for the European Commission (EC) to set up a pilot phase of trading in 2005[24] to gain experience prior to the KP commitment period from 2008 to 2012 and requested the council take actions to introduce flexible mechanisms in to the European Community.[25] A subsequent paper discussed the need to organize a consultative process with stakeholders based on a Green Paper regarding policy options that would need to be considered in the development of an ETS and the potential for developing a pilot phase.[26] The Green Paper was then developed laying out policy options for an ETS and questions to be answered.[27] The EU Emissions Trading Directive establishing a scheme for GHG emissions allowance trading was adopted on 13 October 2003.[28]

---

[22] US Environmental Protection Agency. *Acid Rain Program Benefits Exceed Expectation.* Available from: http://www.epa.gov/capandtrade/documents/benefits.pdf. [No date of publication provided].

[23] US Environmental Protection Agency. *NOx Budget Trading Program/NOx SIP Call, 2003–2008.* Available from: http://www.epa.gov/airmarkets/progsregs/nox/sip.html Last updated 2011.

[24] Commission of the European Communities. *Climate Change—Towards An EU Post Kyoto Strategy.* COM (1998). 353 Final. 03 06 1998. P. 20

[25] Commission of the European Communities, *Climate Change.* P. 21.

[26] Communication From the Commission To the Council And The Parliament. *Preparing for Implementation of the Kyoto Protocol.* COM (1999) 230. 19 May 1999. P. 15.

[27] Commission of the European Communities. *Green Paper on greenhouse gas emissions trading within the European Union.* COM (2000) 87. 8 3 2000.

[28] European Parliament and Council. 2003. Directive 2003/87/EC of the European Parliament and of the Council of 13 October 2003 establishing a scheme for greenhouse gas emissions trading

A brief description of the ETS is necessary given that it was and remains the largest GHG emissions trading program in the world, and it created large-scale demand for CERs. It already has and will continue to influence the development and design of subsequent trading programs around the world in climate change 2.0. The EU ETS initially covered approximately 11,500 installations in the electricity, oil refining, ferrous metals, cement, lime, ceramics, bricks, glass, and pulp and paper sectors across 27 countries once Bulgaria and Romania joined. These facilities emitted approximately 45% of the EU's CO_2 emissions and a little less than 40% of GHG emissions in 2010.[29]

The first phase of the EU ETS ran from 2005 to 2007 and was generally viewed as a pilot phase in which firms would gain experience with the trading mechanism. Each member state was responsible for developing a National Allocation Plan (NAP) that included the number of allowances it planned to allocate and the mechanism for doing so. Phase 2 ran from 2008 to 2012. EU allowances (EUAs) were generally allocated for free in Phases 1 and 2.

The EU adopted a linking directive, which authorized regulated installations to use CERs and ERUs to comply with their EU ETS targets.[30] The directive created demand for CERs and ERUs, and linked the EU ETS to Kyoto, creating the conditions for the global market. CERs and ERUs also became fungible in all Annex I countries. As is discussed in this section, EU firms and governments were the largest purchasers of project-based reductions. Phase 3 of the ETS is currently in place and runs from 2013 to 2020 and is designed to achieve reductions of 21% below 2005 levels from sectors covered by the ETS.[31] Phase 4, which would reduce

---

within the Community and amending Council Directive 96/61/EC. Brussels, European Parliament and Council.

[29] Ellerman, D., F. Convery, C. de Perthius (2010), *Pricing Carbon: the European Union Emissions Trading Scheme*, (New York and Cambridge, Cambridge University Press) P. 28.

[30] European Parliament and Council. 2004. Directive 2004/101/EC of the European Parliament and of the Council of 27 October 2004 amending Directive 2003/87/EC establishing a scheme for greenhouse gas emission allowance trading within the community in respect, of the Kyoto Protocol's project mechanisms. Brussels, European Parliament and Council.

[31] European Parliament and Council. 2009. DIRECTIVE 2009/29/EC OF THE EUROPEAN PARLIAMENT AND OF THE COUNCIL of 23 April 2009 amending Directive 2003/87/EC so as to improve and extend the greenhouse gas emission allowance trading scheme of the Community. Brussels, European Parliament and Council.

trading sectors' GHG emissions 43% below 2005 levels from 2021 to 2030, is working its way through the EU approval process.[32]

Brief analysis is provided later in the section regarding the performance of the ETS in Phase 1.

The Market Responds to Clarity: The EU ETS and Potential for Kyoto

This section illustrates market activity in response to the events cited above. These include the (i) pre-compliance period, which ran until 2004; (ii) the period from 2005 to 2007, corresponding to the first phase of the EU ETS, that created demand for GHG emission reductions from installations regulated by the scheme; and (iii) the entry into force of the KP on 16 February 2005, which would create large-scale demand for GHG emission reductions from 2008 to 2012. However, implementation of firms' and governments' compliance strategies necessarily began earlier, given the lead times required for firms to achieve emissions reductions in their own assets, to identify and contract for CERs that would be usable for compliance in Phase 1 of the ETS, and to identify and contract for CERs and ERUs that could be used during the Kyoto period.

The Pre-compliance Period: 2001–2004

The WB began to publish data on carbon market activity in 2001 based on information provided by Natsource and others.[33] The information provided by Natsource was based on the company's knowledge of brokered transactions in the over-the-counter markets, including deals the company had arranged and media accounts of others.

Because of the progress made in developing the CDM's rules, the anticipated demand for GHG emission reductions that could be created

---

[32] European Commission. Revised emissions trading system will help the EU achieve its climate goals. Specifics regarding the proposal are Available at: http://ec.europa.eu/clima/news/articles/news_2015071501_en.htm [Accessed 13 Janaury 2015].

[33] Rosenzweig, R., D. Forrister. *Natsource Compiles First Comprehensive Analysis of the Greenhouse Gas Trading Market.* [Press Release] 6 August 2001. I have been unable to locate the first State and Trends of the Carbon Markets completed by the PCF in 2001.

by the EU ETS, and potential for the KP to enter into force, carbon market activity increased significantly beginning in 2003. Traded volumes of project-based GHG emission reductions increased from approximately 29 $MtCO_2e$ in 2002 to over 77 $MtCO_2e$ in 2003, and increasing to 107 $MtCO_2e$ in 2004.[34] Purchases for compliance purposes increased from approximately 50 % in 2002 to approximately 90 % in 2003; with nearly all purchases made for compliance in 2004.[35] In the 2003–2004 timeframe, nearly half of the purchases were made by the WB's Carbon Finance Business and the Netherlands, with Japan accounting for 41 %.[36] Beginning in 2004 and through April of 2005, in preparation for the EU ETS, European buyers accounted for 60 % of purchases and Japan accounted for 21 % of purchases.[37]

A small amount of trades of EU allowances also occurred in 2004.

Natsource: Creating a Business

Natsource began to consider the carbon market as a business opportunity in the mid-1990s. This section details the business the company created and the considerations that led to it.

In November of 2000, I joined Natsource to create its Advisory and Research business and to assist the company develop a strategy to participate in the carbon markets. The goal of the research business was to (i) build internal capacity and expertise in the emerging markets; (ii) develop relationships with private firms that would be interested in working with the company in the market; and (iii) to be a stand-alone entity.

By this time, I had been engaged in market-based environmental programs and climate change policy for over a decade. In the private sector, I had participated in the development of the acid rain trading program included in the CAAA of 1990. In government, I was engaged in the creation of the USIJI and played a lead role in creating the climate challenge

---

[34] Lecocq, F., K. Capoor. *State and Trends of the Carbon Market 2005*. International Emissions Trading Association and the World Bank. 2005. P. 20.

[35] Lecocq, F., K. Capoor. *State and Trends 2005*. P. 20.

[36] Lecocq, F. *State and Trends of the Carbon Market 2004*. The World Bank. 2004. P. 20.

[37] Lecocq, F., K. Capoor. *State and Trends 2005*. P. 21.

during the development of the CCAP. Following my departure from government in 1996, I worked with large utilities and energy trade associations in the lead up to the Kyoto negotiations and after to assist them formulate their responses. I also worked with the Environmental Defense Fund and several companies to develop legislation that would have provided firms with credit for reducing GHG emissions prior to there being a legal requirement to do so. The common thread in all of this work was market mechanisms.

I was interested in acquiring commercial experience in the carbon market. Natsource seemed like the perfect place to do this. At this point in time, Natsource was an introductory broker of environmental and energy products in the over-the-counter markets. The company started with a blank sheet in its attempt to create a business to participate in the carbon markets. However, based on his experiences as a broker, Jack Cogen believed it was necessary to develop an alternative business model to brokering. One significant problem with the brokerage model is the transfer of intellectual capital to customers. Brokers in environmental markets typically possess significant knowledge of the rules that govern the markets in which they deal. This knowledge is particularly important to completing transactions in 'quirky', illiquid markets, such as for off-sets, which are created and designed by governments. In order to secure business, brokers share a good deal of knowledge with their customers and, in the process, transfer significant intellectual capital. This is not sustainable. Once the intellectual capital is transferred, smart people replicate it. And when this occurs, there is no need to pay brokers. In addition, once commodity markets become standardized, as they did for SO_2 and NO_X allowances, fees get reduced.

Although Natsource was in the earliest phase of creating a new business strategy, it appeared that the company's new emphasis would be on providing services in the markets for CERs and ERUs. Participation in large CDM or JI project transactions was going to require considerable amounts of capital. Further, the large size of the deals meant intermediaries would be required to provide structuring services such as securing credit, providing project finance, and potentially, syndication. Brokers do not typically possess such skills. The two types of businesses, which had access to capital and structuring expertise, were banks and asset managers.

Since we were not going to become a bank, we decided to build an asset management business. A description is provided later in this section of three of the large CDM transactions Natsource completed and the one that got away. The successful ones all required some of the structuring services referred to above.

Regardless of the path we chose, it was going to be challenging. We were going to need to make significant investments in staff and systems to build the business. Investing customers' money requires an entirely different skill set than introducing counter-parties in the market. Ordinarily, the new investments would have been financed by the cash flow generated by the brokerage business. However, the brokerage business in energy and environmental markets would be forever changed by the meltdown and bankruptcy of Enron in 2001. Enron had been the preeminent player in these markets. Following Enron's demise, trading activity declined greatly, as did Natsource's revenue, limiting the company's ability to finance new business initiatives from cash flow. As a result, Natsource sold off pieces of the declining energy brokerage to raise needed cash.

One other item worthy of note regarding the collapse of Enron: policymakers and much of the public became much more cynical of the ability of markets to solve various public policy challenges. This was exacerbated by the economic decline that began in 2007.

Natsource Carbon Market Initiatives

While developing the new business, Natsource continued as a broker in the carbon market, arranging the first transactions in newly created domestic GHG emissions trading programs. These included the first deals for GHG emissions allowances between DuPont and Mieco, a subsidiary of Marubeni Corporation in the UK's newly created program[38]; the first international trade in GHG emissions allowances under the Danish trading program between Elsam and Entergy Corporation, a US-based power

---

[38] Cormier, L., R. Lowell. DuPont and Marubeni Execute First UK Greenhouse Gas Emissions Allowance Trade. [Press Release]. 21 September 2001.

company that had made a GHG commitment[39]; and, the first swap of government-backed UK and Danish GHG emission allowances between the Royal Dutch/Shell Group of Companies and Elsam SA.[40] Natsource also was involved in arranging transactions for project-based reductions in the pre-compliance markets.

The Emissions Market Development Group

The emission market development group or EMDG, was the first carbon market initiative the company developed. It was a joint venture between Natsource, Swiss Re, Arthur Andersen, and Credit Lyonnais and was launched at The Hague in COP 6 in 2000. Its goal was to create a rating service for project-based GHG emission reductions and to create a diversified portfolio of saleable reductions. It was developed upon two premises. The first was that industrial companies had the ability to create CDM and JI projects within their own assets and second, they would be of some value in the emerging market.

In 2000, the rules regarding the creation of CERs and ERUs were in their infancy and there was no way to estimate their value with any precision. Attempting to estimate CER and ERU's value required an understanding of market activity and the criteria that were guiding project development. We possessed both. Natsource had extensive knowledge of market developments based on its broker's attempts to arrange transactions of GHG emission reductions created by projects. And on the regulatory front, me and my colleague Dirk Forrister, the former Director of President Clinton's Climate Change Task Force and the current President of the International Emissions Trading Association, had both participated in the development of USIJI while serving in government.

---

[39] Pollard, Y., L. Winum. U.S. Utility and Danish Electricity Supplier Conduct First Trade in Danish Greenhouse Gas Allowances. [Press Release].6 December 2001. At the time of this writing, there is no record of this trade on the internet.

[40] Edward, G., L. Winum. Danish Electricity Supplier Elsam Conduct First Ever Transboundary Swap in Greenhouse Gas Compliance Instruments. [Press Release]. 7 May 2002. At the time of this writing, there is no record of this trade on the internet.

The goal of EMDG was to use our regulatory and market expertise to build a model to evaluate and rate emission reduction projects. For example, if the model assigned a project a B rating, we would estimate that it would create 70% of the reductions for which it was designed. Correspondingly, if a project received an A rating, more reductions would be estimated and fewer if it was assigned a C. The goal in all of this was to create a diversified portfolio of saleable—or risk assessed—GHG reductions. The project owners would own a portion of the reduction portfolio and hold an equity position in the company. The service providers would be paid through a combination of reductions and management fee. The initiative was short-lived. Once the rules were agreed to in Marrakesh, the market had no need for a product like EMDG.

Going Global

In recognition that the carbon markets were going to be global, Natsource took steps to increase its presence in Asia, which was going to be important, both on the demand and supply sides of the market. It was anticipated that Japan was going to be a large buyer of CERs and ERUs, and China was going to be an important supplier. Early in 2001, we established Natsource Japan (NJ). It was an outgrowth of the dialogues cited in Chap. 2 that were held between Natsource and Japanese companies that wanted to increase their understanding of environmental markets in the mid- to late 1990s as the international negotiations unfolded. Natsource also had a Japanese investor at the time who was able to introduce us to companies interested in the market. In addition to Natsource, owners of NJ included a mix of trading houses and industrial companies such as Cosmo Oil, Mitsubishi Corporation, Mizhuo Securities, Osaka Gas, Summitomo Corporation, Tokyo Gas, and Toyota Tsusho. Some of these companies would ultimately become participants in Natsource's first carbon fund.

The barriers to creating a presence in Asia to build relationships with buyers and sellers are prohibitive for a company of Natsource's size. NJ was an economically viable way to achieve these goals.

Asset Management

Natsource's first attempt to develop an asset management business in 2001–2002 was called C-Tech. It was modeled after a private equity fund. It would raise capital to invest in companies developing products and services to reduce GHG emissions and create clean energy and the underlying technologies. The effort failed for a simple reason. We did not possess the type of expertise required to succeed in that space. At the same time, our London office, which was established as an electricity and gas broker, had become more involved in GHG markets because of a domestic trading system in the UK and was exploring the development of some type of carbon fund for financial investors.

The company's most prominent asset management effort—the development of the GHG credit aggregation pool (GG-CAP), or buyers' pool, was initiated in 2002. It ultimately became the largest private sector carbon fund in the world. The concept was to pool buyers' demand and use their balance sheets to purchase large volumes of CERs and ERUs while securing favorable prices. It was designed as a turnkey solution. GG-CAP would use Natsource's origination team, consisting of former brokers, to identify projects, and regulatory experts to navigate the project cycle to secure project approvals, contract for the reductions, and manage delivery of them to participant's registry accounts. Unlike C-Tech, Natsource possessed the policy and regulatory expertise and the knowledge of project-based mechanisms and environmental commodities to make GG-CAP work.

We hired a fund manager, Paul Vickers, who had been responsible for TransAlta (TA) Corporation's participation in the carbon market. TA was a large electric utility and one of Canada's largest GHG emitter at the time. The company had done some of the first deals for project-based GHG emission reductions and was a Natsource customer. In his work in the market, Vickers had determined that pooling buyers' demand would be necessary to purchase the output from large projects.

The goal of GG-CAP was to provide buyers with competitively priced CERs and ERUs they could use to comply with their emission reduction obligations. This was not easy. Creating emission reductions from CDM and JI projects was risky and complex. Thus, we expended significant effort in building the delivery risk model (DRM) which could assess the risks in CDM and JI projects and quantify their ability to deliver contracted volumes. It was developed based on years of experience gained in the carbon market, and in cooperation with emissions and commodity traders, risk managers, climate and energy modelers, and technology experts. The DRM incorporated the categories of risk that could affect a project's ability to create volumes of CERs and ERUs. These categories of risk included counter-party risk, the host country's economic policies, domestic and international regulatory risk, and technology risk. It identified events for each risk category that could trigger under-delivery of contracted volumes and developed scoring and weighting assumptions for each. It then utilized various calculations to derive a delivery shortfall for each project. Risk-scoring methodologies were updated continuously based on experience with CDM and JI projects. The model inputs were updated semi-annually to reflect regulatory changes, host countries investment climate and regulatory policies, and experience with CDM and JI projects. The DRM was an important tool in our fund's project evaluation and contracting process.

We designed GG-CAP with several Canadian companies. It is frequently forgotten that Canada was an enthusiastic supporter of market mechanisms in the international negotiations and one of the largest buyers in the pre-compliance market before the country determined that it was not going to take any actions to comply with its Kyoto obligations.

Since GG-CAP was a first-of-its-kind product, there were several challenging issues to work through. It was not a typical investment fund. GG-CAP was structured as an agency agreement. As manager, we would serve as purchasing agent for the participants by using a limited power of attorney to execute contracts on their behalf for a specified volume of CERs and ERUs. In theory, this meant Natsource had an unlimited call on participants' balance sheets. To address this concern, a cap, (known as the maximum variable cost trigger), was included in the management contract on the price we could purchase CERs and ERUs for. This provided participants certainty regarding their maximum outlays, which was the product of their purchase commitment multiplied by the maximum price. It turns out that this adversely impacted the fund's ability to transact. This is because the cap was frequently set at a limit that was below the market price, thus limiting our ability to compete for high-quality projects. An elaborate time consuming process culminating in a 75 % vote by volume was required to increase the maximum price.

Another challenging issue was to fix Natsource's compensation as GG-CAP's manager. In a traditional investment fund, the managers' remuneration consists of a management fee and incentive compensation. The management fee is frequently set at 2 % of committed capital and is supposed to cover operating costs. Incentive compensation is typically 20 % of the fund's profit. Because GG-CAP was not designed to make a profit, a different form of incentive compensation was required to align the interests of Natsource and fund participants. The mechanism to do so required Natsource to purchase a percentage of delivered CERs and ERUs from participants at their cost of acquisition. This provided Natsource with the incentive to purchase the lowest cost CERs and ERUs that could be used for compliance. Natsource would realize its profit by taking the spread between the CER acquisition price and the price we could sell them for in the market. This approach imposed significant market risk on Natsource, which is discussed later.

In the first half of 2004, with continued uncertainty as to whether Russia would ratify the KP, Chugoku Electric became the first participant in GG-CAP. We were off and running.

The Market Takes Off: Kyoto Enters into Force, and Phase 1 of the EU ETS Begins

The major events that were required to jump-start the carbon markets occurred in 2005. Following Russia's ratification on 18 November 2004, the KP took effect on 16 February 2005 requiring 37 developed countries to achieve a fixed level of GHG emissions in the 2008–2012 period from a 1990 base year.[41] Once Russia ratified the KP, the market took off and Natsource's efforts to secure participation in GG-CAP became much easier. We announced a first close of GG-CAP in February 2005 with

---

[41] *Kyoto Protocol.* Article 3 and Annex B. United Nations. 1998.

€72 million (approximately $95 million) committed by six Canadian, European, and Japanese participants, including the Electricity Supply Board of Ireland, The Chugoku Electric Power Company, Hokkaido Electric Power Company, and Osaka and Tokyo Gas Companies.[42] The announcement resulted in a first-page story in the *Wall Street Journal*.[43]

At approximately the same time, Natsource raised additional capital in the form of managed accounts from financial investors to deploy in emissions and renewable energy markets with an emphasis on the carbon market. A managed account was one other than a collective investment vehicle for which Natsource Asset Management LLC had the right and obligation to exercise investment discretion.

And then things got a little crazy. In October 2005, we announced that GG-CAP had grown to €455 million committed by 26 participants from Europe, Japan, and Canada. It ultimately was closed at €510 million. Participants included prominent companies like Endesa Generacion, E.ON UK, and Repsol.[44] The announcement was picked up by press around the world. A few points of interest: one of the items that received the most press attention was that Sergey Brin, the President and Co-Founder of Google participated in the fund. The other more substantive item is that GG-CAP participants were a mix of small and large companies. Our thinking in designing the fund was that small- and mid-sized companies were more in need of a product like GG-CAP than large firms. This was because small power companies with GHG compliance obligations did not have the resources to originate CDM projects in China, assess them, and move them through the project cycle. Their business was to provide kilowatt hours to their customers. As such, GG-CAP was designed as a turnkey solution for such firms. In contrast, our view was that large companies would build the internal capacity to originate, evaluate, and contract CDM and JI projects as an element of a diversified GHG compliance strategy. It turned out that large companies were

---

[42] Rosenzweig, R. Natsource Announces Launch of the Greenhouse Gas Credit Aggregation Pool. [Press Release]. 2005.

[43] J. Fialka. Natsource Forms Investment Pool To Meet Greenhouse-Gas Credits. Wall Street Journal. 28 February 2005. P. 1.

[44] Rosenzweig, R. Natsource Closes Greenhouse Gas Credit Aggregation Pool. 2005. [Press Release]. 2005

so in need of GHG emission reductions they could use to meet compliance obligations, that they joined funds like GG-CAP. Because of this, GG-CAP sometimes competed with fund participants for deals.

Natsource was not the only carbon specialist company to experience rapid growth. EcoSecurities, Trading Emissions PLC, and Carbon Asset Management Company went public on the AIM market on the London Stock Exchange in 2005. AgCert International went public on the London Stock Exchange. Although these companies' business models were slightly different, they all focused predominantly on carbon, with an emphasis on the project markets. Investors were bullish on carbon.

The increased interest in carbon created by entry into force of the KP and Phase 1 of the EU ETS resulted in rapid growth in volume and value for both EUAs and CERs and ERUs.

What follows is a brief summary of market activity from 2005 to 2007.

Phase 1 of the EU ETS: 2005–2007

The goal of this section is to provide a brief review of market activity in Phase 1 of the EU ETS and its performance.

The performance of Phase 1 was mixed. Trading activity increased significantly in each of its three years of operation. Volumes of EUAs transacted increased from 8.49 million in 2004 to over 322 million in 2005 and value increased to nearly $8 billion.[45] Growth continued in 2006 with traded volumes of EUAs increasing to over 1.1 billion. Traded volumes of EUAs increased to over 2 billion in 2007 at a value of nearly $50 billion, although most of the value was based on transactions of Phase 2 allowances.[46,47]

---

[45] Capoor, K., P. Ambrosi. *State and Trends of the Carbon Market 2006.* The World Bank and the International Emissions Trading Association. 2006. P. i.

[46] Capoor, K., P. Ambrosi. *State and Trends of the Carbon Market 2008.* The World Bank. 2008. PP. 1–2.

[47] It is important to note that the EU shows lower volumes of traded EUAs during this time period from what is included above in another document. I used the higher volumes although both sets of data show growth in traded volumes of EUAs. The lower volumes can be found at: http://ec.europa.eu/clima/publications/docs/factsheet_ets_en.pdf

Phase 1 of the EU ETS will be remembered for two events. The first was a price decline in EUAs from approximately €30 in April of 2006 to €10 in the next 60 days. The rapid price decline was caused by the release of verified emissions data to the market confirming several EU members had surplus EUAs and would not be required to make additional emission reductions. During the first round, member states made generous initial allocations to regulated installations based on data provided by companies. This resulted from a lack of data and the minimal amount of time to get the system up and running. In addition, although prices for Phase 1 EUAs recovered slightly in the next few months from €10, they ultimately lost all of their value in 2007 because of a prohibition on banking surplus Phase 1 EUAs into Phase 2.[48] Because many installations were already in compliance with their Phase 1 limits, they had no further use for EUAs in Phase 1, rendering them valueless.

In assessing Phase 1 of the EU ETS, the two events that will be remembered, the precipitous price decline of EUAs and their fall to zero, were both caused by design elements. Government's design of environmental markets has always and will continue to impact their performance. However, because Phase 1 was viewed as a pilot and was prior to the KP commitment period, it is not necessary to spend significantly more time assessing its performance.

Chapter 4 provides a more thorough review and analysis of the performance of Phase 2 of the EU ETS. It attempts to answer the important questions of whether it achieved emission reductions and stimulated investment in the low and non-emitting technologies that will be required to achieve long-term climate policy objectives. It is important to take both a retrospective look at the EU ETS because of its importance in creating the global carbon market in climate change 1.0 and the influence it will likely continue to have on future policy-making.

---

[48] Capoor, K., P. Ambrosi. *State and Trends of the Carbon Market 2007*. The World Bank and the International Emissions Trading Association, 2007. P. 12.

CER and ERU Market 2005–2007

This section illustrates some of the characteristics of the CDM market during this period including contracted volumes of CERs, dynamics that impacted their pricing and some controversies that emerged during this period.

CER Volumes

The market for CERs grew significantly during 2005–2007 because of the demand created by the EU ETS and entry into force of the KP. Transacted volumes of CERs increased from approximately 97 $MtCO_2$e in 2004 valued at approximately $485 million,[49] to 350 $MtCO_2$e in 2005 valued at approximately $2.6 billion,[50] and to approximately 560 $MtCO_2$e in 2006 valued at $6.250 billion.[51] And finally, the market increased to nearly 800 $MtCO_2$e valued at approximately $12.8 billion in 2007.[52]

Until 2007, the large majority of transacted volumes were for primary CERs, which for purposes of this book, are defined as CERs that have not yet been issued. In 2007, a significant secondary CER market emerged. These are contracts in which the seller provided some type of delivery guarantee or sold an issued CER. The market for secondary CERs increased to 240 $MtCO_2$e (approximately 30% of transacted CERs) in 2007 valued at approximately $5.4 billion, representing over 40% of the CERs' market value.[53] JI was inconsequential during this period always accounting for less than 10% of the transacted volumes.[54]

Some Dynamics Affecting CER Prices

CER prices were influenced by several variables during this period. Some of these included the project's status in the project cycle, the allocation

---

[49] Capoor, K., P. Ambrosi. *State and Trends 2006.* P. 23.
[50] Capoor, K., P. Ambrosi. *State and Trends 2007.* P. 20.
[51] Capoor, K., P. Ambrosi. *State and Trends 2008.* P. 1.
[52] Capoor, K., P. Ambrosi. *State and Trends 2008.* P. 1.
[53] Capoor, K., P. Ambrosi. *State and Trends 2008.* P. 1.
[54] Capoor,K., P. Ambrosi. *State and Trends 2008.* P. 1.

of risk between the counter-parties, the creditworthiness of the seller, whether the buyer assisted the seller navigate the project cycle or provided project finance, and China.

Regarding the project cycle, higher prices were paid if contracts were completed as the underlying project moved closer to registration or verification. The seller could also get a higher price by providing some form of delivery guarantees and if it was a creditworthy entity. If the buyer assisted the seller develop PDDs, secure registration, or provided some type of project finance in the form of debt or upfront cash for the purchase of equipment necessary to implement the project activity, lower prices were paid.

China also had a great influence on CER prices because it was hosting projects that created the large majority of transacted volumes. The Chinese DNA would not approve a contract unless the developer received a certain price. This became an unofficial floor price in the market.

Buyers and Sellers

Other parts of the market remained fairly constant from 2005 to 2007. The largest buyers of CERs were the EU, representing 50% of transacted volumes in 2005[55] and increasing to nearly 90% in 2007.[56] Japan was also a consistent buyer, but its portion shrank over time. China was the dominant supplier.[57]

Controversies Surrounding Industrial Gas Projects

Industrial gas projects that reduced hydrofluorocarbon (HFC)-23 and nitrous oxide (N2O) were popular during the early period of the CDM. HFC-23 was responsible for 67% of transacted volumes in 2005 and 34% in 2006. N20 captured a 13% share in 2006.[58] They were also

---

[55] Capoor, K., P. Ambrosi. *State and Trends 2007*. P. 22.

[56] Capoor, K., P. Ambrosi. *State and Trends 2008*. P. 23.

[57] Capoor, K., P. Ambrosi. *State and Trends 2008*. P. 26.

[58] Capoor, K., P. Ambrosi. *State and Trends 2007*. P. 27.

controversial, particularly HFC-23 destruction projects. A general review of what occurred follows.

Some argued that the CDM provided an incentive to ramp up production of HFC-22, which was scheduled to be phased out under the Montreal Protocol on Substances That Deplete The Ozone Layer in 2020.[59] This is because increased production of HFC-22 created additional HFC-23, which was not controlled under the Protocol. Since HFCs were regulated under the KP, CDM projects could be developed to destroy them and PPs could earn money for doing so. This opportunity for revenue generation was exacerbated by several factors. These included HFC-23's high global warming potential and large producing factories resulting in the creation of significant volumes of CERs. I am not going to argue that firms did not participate in gaming or excuse it.[60] However, firms gaming programs to maximize revenue is not a new phenomenon. It happens whenever money is at stake.

A few alternative points of view follow regarding the benefit of HFC projects in the CDM.

1. *HFC-23 emissions were not controlled*—The bottom line is that these emissions were uncontrolled. The CDM provided a financial incentive to eliminate them. Without it, HFCs would have continued to be freely vented in to the atmosphere.

2. *HFC-23 projects were additional*—This is one of the few areas of agreement. These emissions were reduced from what they would have been without the project activity.

3. *Industrial gas projects kept prices low*—The volumes of CERs created by industrial gas projects were critical to building confidence in the CDM and the carbon market in its earliest period. In the absence of CERs created by industrial gas projects, supplies would have been limited, potentially putting upward pressure on prices. This would have caused a problem at the beginning of the effort to address climate change.

---

[59] United Nations Environmental Programme. *The Montreal Protocol On Substances That Deplete The Ozone Layer.* United Nations. 1987.

[60] Wara, M. (2008). 'Measuring The Clean Development Mechanism's Performance and Potential', 55 *UCLA Law Review* (2008). 1781–1789.

4. *HFC-23 and N2O reduction projects did not crowd out efficiency and renewable energy projects*—Opponents of industrial gas projects argued that they crowded out investment in renewable energy and energy efficiency projects. This is not the case. The CDM worked as economists concluded it should. HFC and N20 projects were among the first implemented because their abatement costs were low and they were less risky than other CDM project types. Regulatory risk was low because the first approved methodology in the CDM was for HFC-23. And counter-party risk and technology risks were low. Once industrial gas projects were exhausted, transacted volumes from energy efficiency and renewable energy projects increased from 14 % in 2005 to 64 % in 2007.[61]

The controversy over HFC, and to a lesser degree N20 projects, damaged the credibility of the CDM, reducing its political viability. Policies such as the CDM cannot succeed without the support of affected parties. I do not believe the CDM ever recovered from this and other issues. This dynamic is discussed in greater detail in Chap. 4.

Major problems began to surface in the CDM during this period. The mechanism came under intense criticism from developers due to bottlenecks in the project cycle and others who believed it was providing CERs for activities that were not additional. These issues are the subject of discussion and analysis in Chap. 4. The arguments regarding the CDM would be similar to those regarding any project-based offset system and have shaped my views regarding their future role in climate policy.

Natsource Becomes the Largest Buyer of Contracted CERs

From 2005 to 2007, Natsource completed three very large CDM transactions. And other large deals got away. Brief descriptions of these transactions are provided to illustrate Natsource's participation in the market, its evolution from a broker to asset manager, and the EBs performance.

---

[61] Capoor, K., P. Ambrosi. *State and Trends 2008.* P. 28.

The World Bank Umbrella Carbon Facility

The WB UCF contracted to purchase more than 129 million CERs from 2006 through 2013 from HFC-23 incineration projects located at two manufacturing facilities in Jiangsu Province in China. The bank valued this deal at over $1 billion, the first and potentially only billion dollar CDM deal ever completed. Members of the UCF included five carbon funds and 11 private sector firms. Some of the private firms included financial houses Deutsche Bank and Mitsui and Co., large power companies Endesa, RWE, and Tokyo Electric Power Company, and carbon specialist companies Natsource, Climate Change Capital, and Trading Emissions PLC.[62]

Natsource was the largest private sector purchaser in the deal, which closed in August 2006, contracting to buy 23 million CERs. It established two special purpose entities, Canadenis Acquisition Limited and Tamarisk Acquisition Corporation to participate in the deal. Canadenis was created for GG-CAP participants and Tamarisk was created for our financial investors to participate in the transaction.[63]

It is an interesting story how two Natsource entities were able to participate in the UCF. We were informed that the first ten applications received by fax at the bank would be in the deal. In preparation for the submissions, Mike Grande, our IT director, calculated that it would take 18.5 seconds for faxes sent from our Exchange email server to reach the WB. In order to avoid a busy signal on our fax line during the redial process, he programmed a 0.5 second delay between sending each fax. At exactly 18.5 seconds before the time applications would be accepted, he began to fax them. One by one we watched the server churn out multiple copies of our applications. With each successful transmission, we knew Natsource would have to be at least one of the first ten accepted. Twenty minutes later, Jack Cogen received a call from the WB. He was told that seven of the first ten applications received were from our two funds. The bank had also asked that we stop bombarding their servers!

---

[62] The World Bank. *Umbrella Carbon Facility Completes Allocation of First Tranche.* [Press Release] 30 August 2006.

[63] Rosenzweig, R. *Natsource Announces Participation in the Largest Greenhouse Gas Transaction on Record.* [Press Release] 29 August 2006.

The UCF transaction was particularly difficult for Canadenis to complete because of GG-CAP's design and structure. A few examples illustrating the complexities follow. Several GG-CAP participants opted to not participate. Because of this, we scrambled to get some of the remaining participants to agree to purchase the resulting excess CERs. This altered the proportionate shares of GG-CAP participants' aggregate purchase commitments. Since all deals and expenses were allocated based on this principle, this had to be readjusted. We also needed participants to agree to buy 2013 vintage CERs, even though most were prohibited from so doing because of the uncertainty regarding the post-2012 period. GG-CAP was also prohibited from purchasing post-2012 CERs. In order to make this work, we arranged for a swap among participants with some agreeing to take the 2013 vintages and others taking a higher percentage of earlier vintages.

From a structuring perspective, a letter of credit (LC) was required to secure Canadenis' purchasing obligations. This was the first time Natsource was required to do this. This was made more challenging because the provider of the LC had little experience in the carbon market and required significant education. In addition, once they ultimately decided to provide the LC, they required each of the participants to secure 110 % of their obligations to guard against the potential default of others. Participants did not like this provision, as it required them to obligate a larger percentage of their companies' capital to the transaction. Also, at the last minute, the LC provider required the companies that had agreed to purchase the excess CERs created by the opt-outs to provide a letter committing them to secure their obligations in the amount of the value of the additional CERs they had purchased. Although this seemed to be a reasonable request, it was made less than a week prior to the scheduled deal closing. This required us to locate representatives from companies authorized to make such commitments at the end of August, which is typically vacation time.

We were excited to complete this transaction for several reasons. Our approach in pooling a large amount of capital had been vindicated and put us on the map in the market and in China. The UCF deal allowed us to meet a large volume of GG-CAP's volumetric commitment and was potentially lucrative. Based on GG-CAPs performance compensation provision described previously, the UCF would provide the company with hundreds of thousands of CERs to monetize in the market at a significant spread. We would also realize a large profit from monetizing the CERs in the market for the managed accounts and hedge funds' financial investors. Of course, given the subsequent crash in CER prices, the economics did not work out as we hoped. This too is described in Chap. 4.

Natsource Completes CDM Transaction with Henan Shenma Chemical Company

Natsource's next large transaction was an N2O decomposition project in China. The project activity was the installation of a catalyst to reduce N2O emissions from adipic acid production. It was designed to reduce GHG emissions by approximately four $MtCO_2e$ per year.[64] At the time of its completion in 2006, Shenma was the 13th largest CDM transaction. We were able to secure this deal for several reasons. These included the reputation the company established in the UCF transaction and relationships that the company had with the catalyst manufacturer from prior transactions in the market.

This was a complicated and innovative deal for 23 million CERs. The deal could only be completed because we had industrial participants in GG-CAP that required the CERs for compliance, the managed accounts, and other funds with different objectives. Shenma required capital to purchase the catalyst necessary to implement the project. Our financial investors were able to provide the capital. In contrast, GG-CAP participants would not take such risk but had the wherewithal and desire to purchase large CER volumes at fixed prices. The Shenma transaction represented the first time Natsource provided project finance. The capital providers were paid off with the first CERs from the project.

Pricing was another innovative component of the deal. At this point in time, prices for primary CERs were increasing. Because of this, sellers were hesitant to lock themselves into fixed price deals. They were seeking

---

[64] For more information on the project see: http://cdm.unfccc.int/Projects/DB/DNV-CUK1176373789.59/view?cp=1

to capture the benefits of higher prices. Yet, GG-CAP was prohibited from transacting above a certain level or from engaging in alternative pricing structures such as indexes or variable or floating prices that were increasing in popularity. In an attempt to accommodate the counterparties' needs, GG-CAP would purchase contracted volumes at a fixed price. This assured the seller of a minimum amount of revenue. The financial investors would have the right to purchase the remaining 50 % of volume at a variable price providing the seller an opportunity to share in some of the upside.

The financial investors had a call option on 50 % of the projects' CERs: they had the right but not the obligation, to purchase those CERs. The GG-CAP buyers had the right and obligation to purchase 50 % of the project's output and the obligation to purchase an additional amount, up to 100 % of the project's CERs if the seller required this.

Natsource Purchases 90 % of the Volumes from PetroChina Company Limited Liaoyang Petrochemical Company CDM Project

In its largest transaction to date, Natsource entered into a contract in 2007 to purchase 90 % of the CERs that would be created by a N2O reduction project implemented by PetroChina Company Limited Liaoyang Petrochemical Company (LYPC), which was the third largest CDM deal ever completed.[65] Natsource was able to secure this transaction because of the track record we had established in China and with N2O projects. Goldman Sachs International purchased the remaining 10 % of the CERs. The project was designed to achieve reductions of more than 10 $MtCO_2e$ per year.

In the previous two transactions, Natsource had acquired experience in arranging LCs, and in providing project finance. Because of the size of this project and because the price per CER exceeded GG-CAP's maximum contract price, we would need to develop a sizable syndicate to complete the deal. A Natsource subsidiary was the purchaser of LYPC's CERs. We

---

[65] For more information on the project see: http://cdm.unfccc.int/Projects/DB/DNV-CUK1184240745.87/view?cp=1

agreed to on-sell the CERs to nine entities, including major financial institutions and entities advised by Natsource Asset Management. As purchaser, we needed an LC to secure our purchasing obligations. Each of the on-sale-buyers was required to secure its percentage of its purchase obligations. Some provided what was essentially cash collateral; others, including major international financial institutions secured their obligations with individual LCs.

There was a humorous situation in which the bank providing the LC to Natsource to cover the entire transaction sent a representative to China with the LC in a briefcase that was handcuffed to his wrist. He and Martin Collins, the director of Natsource origination and a key person in this deal, showed up unannounced to LYPC. They would provide the LC at the same time we would receive the executed ERPA.

The financial crisis created a bizarre situation in this deal. The syndicate had a conference call as we were approaching a delivery date for CERs during the crisis. Because of the way in which the CDM worked and the deal structure, only three entities (Fortis Bank, Goldman Sachs, and Natsource) were registered to take delivery of issued CERs. For our part of the transaction, we used Fortis for credit enhancement. The process was for our syndicate CERs to be sent to Fortis' bank account, which would then distribute them to syndicate buyers following payment. Because of the impact of the financial crisis, there was a risk Fortis could not implement its obligations to settle the deal. If it could not, the syndicate would be in jeopardy of defaulting to LYPC. Because the deal was profitable, no one wanted this to occur. On a call, Jack Cogen suggested that we contact Goldman to see if they could handle the mechanics for us. Someone on the call said that perhaps Natsource should step in because Goldman could also be at risk. Jack asked everyone to pause in order to recognize that we had reached a point in time in which Natsource credit may have been more acceptable than that of a large financial institution. We began to call this the 'too small to fail' moment.

The bottom line was that this transaction was successful for the environment reducing 60 $MtCO_2e$ and provided economic benefits to Natsource.

The Atlantic Methanol Project: One that Got Away

The Atlantic Methanol Project Company (AMPCO) deal was a large-scale infrastructure project undertaken in Equatorial Guinea during the AIJ pilot phase. Prior to joining Natsource in 2000, I had worked with the company applying to qualify AMPCO as a USIJI project. The project constructed a methanol production facility that used previously flared natural gas from offshore oil production as a feedstock to produce methanol.

As described previously, Parties would not earn any credits for 'GHG emissions reduced or sequestered during the Pilot Phase'.[66] Because the KP authorized the use of 2000–2007 vintage CERs for compliance in the first commitment period,[67] developers of projects in the pilot phase wanted to be able to monetize the reductions that were created. However, there was no policy allowing for this to occur.

This changed in 2005 due to a provision within the 2005 Montreal Declaration at COP/MOP-1 and at a subsequent meeting.[68,69] These decisions provided projects that started between 1 January 2000 and 18 November 2004 with the ability to secure CERs if they were registered by the EB on 31 March 2007, providing the project with approximately two years to secure registration. Given our prior relationship with the project owners and CDM expertise, the decision gave Natsource the opportunity to work with AMPCO to attempt to convert millions of tons of otherwise worthless reductions into CERs that were trading for $17/ton at the time.

Natsource moved quickly to communicate to the owners that there was a possibility to convert the project's historic reductions to CERs. We would not tell them how until we entered into a business relationship.

[66] *Report Of The Conference Of The Parties On Its First Session, Held At Berlin From 28 March To 7 April 1995*, Decision 5/CP.1. United Nations. 1995.

[67] *Kyoto Protocol to the United Nations Framework Convention on Climate Change.* Article 12 10. United Nations. 1998.

[68] *Report of the Conference of the Parties serving as the meeting of the Parties to the Kyoto Protocol on its first session, held at Montreal from 28 November to 10 December 2005. Further guidance related to the Clean Development Mechanism.* Decision 7/CMP.1. United Nations. 2006.

[69] *Report of the Conference of the Parties serving as the meeting of the Parties to the Kyoto Protocol on its second session, held at Nairobi from 6 to 17 November 2006. Further guidance relating to the Clean Development Mechanism.* Decision 1/CMP.2. United Nations. 2007.

Natsource agreed to a deal in which we would pay all of the expenses necessary to undertake the work and accept the risk of securing the project's registration in exchange for a percentage of CERs issued to the project.

The first task was to develop and submit a PDD and methodology for quantifying the GHG emissions reductions resulting from closing the flare and using the previously flared gas as feedstock. In theory, this should have been simple. But, like most issues in the CDM, the methodology approval process was adversarial and conducted through intermediaries. The bottom line is we got the methodology approved but needed to overcome a near disaster. The staff in charge of setting the EB agenda left our methodology off the docket for the next meeting. This would have delayed our hearing by 45 days and run out the clock, even though we had submitted the necessary information on time. We were able to fix this.

Having cleared that hurdle, the next step was to revise the PDD in line with the approved methodology and secure validation. Once AMPCO accomplished this—the final step was registration. The vast majority of projects at this time were registered without review—the theory being that DOEs are credentialed to make this determination.

The EB requested a review of the project to determine if it should be registered. The issue was language regarding prompt start project eligibility.[70] It held that only projects that 'start after 1, January, 2000' are eligible to be registered. This would be decided by the words 'Project Start Date', which was a total gray area. Did this mean the project idea, the feasibility study, beginning of construction, or when the emissions reductions started? The project was under construction prior to 2000 but began operation in 2000. Based upon a cursory review of registered projects, it was apparent that nearly all had started in some form prior to 2000. Critical documents, such as government permits, feasibility studies, and construction plans from the 1980s and 1990s were routinely submitted to the EB in support of PDDs.

---

[70] UNFCCC Secretariat. *Report of the 32nd Meeting of the Executive Board of the Clean Development Mechanism, Annex 31.* UNFCCC. CDM-EB-32. 2007.

We would wait the 60 days that the EB was provided to undertake the review. During this time, we also sought to quantify the CER volumes that would be issued once the project was registered. It was one thing to achieve validation and another to have verified reductions from a project. Recognizing the challenges in retroactively documenting emission reductions prior to the development of a new methodology, we attempted to ensure that the methodology did not require any records, documents, or operating parameters that the project would not have on file. And, of course there was always the risk that the EB would unilaterally and retroactively change the approved methodology, disqualifying AMPCO. This had already happened to AgCert, which significantly damaged its prospects.

We needed to verify the reductions against our methodology before the EB could change it. This required hiring a DOE to conduct an onsite verification in Equatorial Guinea, which is not typically viewed as a hospitable country. After a site visit and reviewing data and records, the DOE determined that, once registered, the project would be eligible for issuance of approximately 10 million CERs valued at $170 million.

Securing the CERs was entirely dependent on the EB's decision. It rejected the project based on the start date.[71] And just like that, $170 million in value disappeared. A year later, the EB changed its interpretation of the issue.[72] Under the new policy, it would allow registration of projects that had commenced construction before 2000, were stopped, and then were completed and brought into operation after 2000. This was similar to what occurred with AMPCO, but there was no appeal process or ability to revisit the decision. The arbitrary nature of the EB's decision-making and lack of an appeals process will be discussed in Chap. 5 in the context of the CDM's performance.

In addition to the three successful transactions above, accounting for approximately 100 million CERs, Natsource funds entered into several others on behalf of its fund and managed account

[71] UNFCCC Secretariat. *Report of the 33rd Meeting of the Executive Board of the Clean Development Mechanism.* UNFCCC. CDM-EB-33. 2007.

[72] UNFCCC Secretariat. *Report of the 41st Meeting of the Executive Board of the Clean Development Mechanism, Annex 45.* UNFCCC. CDM-EB-41. 2008.

participants. This was an exceptionally difficult period for the company. We experienced many of the same challenges as other small undercapitalized businesses operating in new markets undergoing rapid growth. We were in the process of acquiring staff with skills better suited to asset management than brokerage, required a stronger administrative infrastructure and additional capital to take advantage of increasing market opportunities and to operate the business. We were attempting to address these challenges and implement the business at the same time.

The prior sections provided a brief review of the rules that increased the certainty to implement CDM projects, and the establishment of the EU ETS and briefly summarized market activity from 2005 to 2007. Chapter 4 assesses the performance of the EU ETS and the CDM from 2008 to 2012.

Back in the US

The US took a back seat while the rest of the world took consequential actions on climate change from 2000 to 2007. However, within the US, activities at the state level and the legislative and judicial branches did have an influence on climate change policies in the first generation of policy-making. More importantly, as is described in Chaps. 5 and 6, they greatly influenced the development of policy in climate change 2.0 and continue to do so. The State initiatives helped to usher in the era of bottom-up policy-making which is currently a foundation of climate change policy at all levels and is now being emulated around the globe. The legislative proposals at the federal level and the debate surrounding them influenced the failed effort to pass a cap-and-trade bill in 2009–2010, bringing climate change 1.0 to an end. And a 2007 Supreme Court decision essentially requiring the US Environmental Protection Agency (EPA) to regulate GHG emissions from motor vehicles established the foundation for the far-reaching regulations that President Obama has put in place and which are the cornerstone of climate change policy 2.0 in the US.

State Initiatives

A brief description of the two most prominent bottom-up policy-making efforts in the US follow.

The Regional Greenhouse Gas Initiative

Seven Northeastern and Mid-Atlantic States signed a memorandum of understanding in 2005 to develop the Regional Greenhouse Gas Initiative (RGGI), a cap-and trade program to limit CO_2 emissions from power plants.[73] In 2015, nine states were participating in the program. Unhappy with the outcome of the CAAA of 1990, participants in RGGI hoped to shape federal climate change legislation they believed to be inevitable by proactively developing their own program. They hoped that the federal government would borrow some of RGGIs key provisions and include them in federal legislation in order to benefit their state and region. Some believed the program would ultimately be subsumed by a federal GHG control program and never take effect.

Launched in October of 2008, and governed by a model rule,[74] RGGI represented the first large-scale effort in the US to address climate change. The program included an aggregate emissions budget for the region representing the quantity of CO_2 that could be emitted. The overall budget was then allocated to States. The first budget period ran from 2009 to 2014, and was set to stabilize power plant CO_2 emissions. The second period ran from 2015 to 2018 and was designed to achieve a 10% reduction.[75]

A review of the program undertaken in 2012 confirmed that RGGI was over-allocated from the program's inception. Emissions were never above the cap. Allowance prices declined from a high of over $3.50[76]

---

[73] Regional Greenhouse Gas Initiative. *Memorandum of Understanding*. Available from: http://rggi.org/docs/mou_final_12_20_05.pdf [Accessed 4 December 2015].

[74] Regional Greenhouse Gas Initiative. *Model Rule*. Available from: https://www.rggi.org/docs/model_rule_8_15_06.pdf [Accessed 4 December 2015].

[75] Regional Greenhouse Gas Initiative. *Overview of RGGI CO₂ Budget Trading Program*. Available from: http://www.rggi.org/docs/program_summary_10_07.pdf [Accessed 4 December 2015].

[76] Regional Greenhouse Gas Initiative. *Auction 3 Results*. Available from: https://www.rggi.org/market/co2_auctions/results/auctions-1-28 [Accessed 4 December 2015].

early in the program, to a level that was at or near a reserve price of less than $2.00[77] in auctions held at the end of the 2009–2011 period. It is likely that lower electricity load resulting from weather, efficiency, the recession, and increased use of nuclear energy and natural gas for power generation contributed to lower CO_2 emissions. In practice, RGGI operated as a small tax on fossil-fueled generation that provided the States with revenue from allowance sales at auction that could be used for investment in energy efficiency, renewables, and other favored activities.

These results, combined with the failure to pass comprehensive federal climate change legislation led to significant changes in the program that were included in an updated model rule that took effect in 2014. The key changes were to reduce the budget by 45 % in 2014, with further reductions of 2.5 % per year from 2015 to 2020.[78] Supply was also reduced by approximately 140 million allowances that States could sell at auctions from 2014 to 2020.[79,80] The impact of the new rule was to turn a program that had been structurally oversupplied since its inception into a market that is structurally short from 2014 to 2020.

The Global Warming Solutions Act

Following RGGI's lead, California enacted the Global Warming Solutions Act (AB 32) in 2006.[81] The legislation required the State's Air Resources Board (ARB) to 'determine what the statewide greenhouse gas emissions level was in 1990, and approve in a public hearing, a statewide green-

---

[77] Regional Greenhouse Gas Initiative. *Auction 14 Results*. Available from: https://www.rggi.org/market/co2_auctions/results/auctions-1-28 [Accessed 4 December 2015].

[78] Regional Greenhouse Gas Initiative. *Summary of RGGI Model Rule Changes*. Available from: https://www.rggi.org/docs/ProgramReview/_FinalProgramReviewMaterials/Model_Rule_Summary.pdf [Accessed 4 December 2015].

[79] Regional Greenhouse Gas Initiative. *First Control Period Interim Adjustment for Banked Allowances Announcement*. Available from: https://www.rggi.org/docs/FCPIABA.pdf [Accessed 4 December 2015].

[80] Regional Greenhouse Gas Initiative. *Second Control Period Interim Adjustment for Banked Allowances Announcement*. Available from: https://www.rggi.org/docs/SCPIABA.pdf [Accessed 4 December 2015].

[81] Nunez, F., F. Pavley. *California Global Warming Solutions Act of 2006*. Available from: http://www.leginfo.ca.gov/pub/05-06/bill/asm/ab_0001-0050/ab_32_bill_20060831_enrolled.html [Accessed 4 December 2015].

house gas emissions limit that is equivalent to that level, to be achieved in 2020'.[82] The intent of the legislature was that the emissions limits would continue past 2020.[83] On 29 April 2015, Governor Brown signed an executive order establishing a GHG reduction target of 40% below 1990 levels by 2030.[84]

The legislation required ARB to develop and approve a Scoping Plan by 1 January 2009 that would achieve the maximum technologically feasible and cost-effective reductions in GHG emissions. 'The plan shall identify and make recommendations on direct emission reduction measures, alternative compliance mechanisms, market based compliance mechanisms.'[85]

In preparing the plan, the ARB was required to 'evaluate the total potential costs and total potential economic and non-economic benefits of the plan for reducing greenhouse gases to California's economy, environment, and public health using the best available economic models, emissions estimation techniques, and other scientific methods'.[86] In adopting the regulations to implement the policies included in the plan, ARB was required to '[c]onsider other societal benefits, including reductions in other air pollutants, diversification of energy sources and other benefits to the economy, environment and public health.'[87]

The compliance plan included an economy-wide cap-and-trade system and numerous CPs, many of which were traditional regulatory measures. These policies were primarily designed to: (i) increase the use of renewable energy sources in the State; (ii) reduce GHG emissions from the transportation sector and the carbon intensity of transportation fuel; and (iii) improve the energy efficiency of the State's economy.

In 2011, ARB determined that if CPs (including aggressive light-duty vehicle GHG performance standards) achieved their estimated emission reductions, they would account for over 75% of the 80 Mt of emis-

---

[82] Nunez, F., F. Pavley. *Global Warming 2006*. Section 38550.

[83] Nunez, N., F. Pavley. *Global Warming 2006*. Section 38551(b).

[84] Brown, Governor E. G. Jr. (2015) *Executive Order B-30-15*. Available from: https://www.gov. ca.gov/news.php?id=18938 [Accessed 4 December 2015].

[85] Nunez, F., F. Pavley. *Global Warming 2006*. Section 38561(b).

[86] Nunez, F., F. Pavley. *Global Warming 2006*. Section 38561(d).

[87] Nunez, F., F. Pavley. *Global Warming 2006*. Section 38562(b)(6).

sion reductions necessary to achieve the legislations target of 1990 GHG emission levels in 2020. In this scenario, the cap-and-trade program would account for just over 20% of reductions.[88] Based on a subsequent revision of the State's 2020 BAU forecast and its 2020 GHG emissions limit, the CPs were estimated to achieve approximately 70% of the GHG emission reductions required for compliance.[89]

The requirement that ARB consider 'other societal benefits' in adopting the plan is important. It is in recognition that in addition to using market-based compliance mechanisms to control the costs of meeting the GHG emissions target, California's policy-makers placed equal importance in achieving other economic, environmental, and public health benefits that many believe are outside of the carbon markets' reach. The CPs were included in the plan to provide those benefits.

In theory, combining market-based policies and regulatory measures makes sense. Although a well-designed cap-and-trade program will stimulate investment in renewables and in improving the efficiency of energy use, other measures are also required to achieve such objectives. In practice, government policies designed to achieve market efficiency and cost control may compete with others put in place to increase the use of renewables and improve energy efficiency.

The EU ETS, and indeed most market mechanisms that are in place to reduce GHG emissions, operate alongside a multitude of other regulatory measures or CPs. Chapter 4 reviews research detailing the impact of such policies on the EU ETS market. Many other jurisdictions developing responses to climate change will employ similar approaches. Because this policy model is increasing in prominence, additional research would assist policy-makers to understand the interactions between carbon markets and CPs and to develop policies that had the best chance of achieving their objectives at the lowest cost. It is important to understand how the regulatory measures will impact the market's ability to operate effectively.

RGGI and California influenced the ongoing era of bottom-up policy-making that is currently underway including the 60 GHG trading and

---

[88] California Air Resources Board. *Status of Scoping Plan Recommended Measures.* 2011 July.
[89] California Air Resources Board. *First Update to the Climate Change Scoping Plan, Building on the Framework Pursuant to AB 32, The Global Warming Solutions Act.* California PP. 92–93.

tax programs in operation or under development at the national, regional, and subregional levels.[90]

Federal Legislative Proposals

We will briefly return to President Bush's early decision to renege on his campaign pledge to regulate power plants emissions of SO_2, NO_X, mercury (Hg), and CO_2. This approach was known as the multi-pollutant approach or 4p by some. It resulted from the power sector's concerns that implementation of the CAA requiring reductions of SO_2 and NO_X, anticipated requirements to reduce Hg emissions for the first time, and uncertainty regarding the future regulation of CO_2 made it difficult to plan their investments in new power plants and to develop efficient pollution control strategies. There would be the potential that uncoordinated, piecemeal requirements to control individual pollutants at varying points in time was inefficient and could result in stranded investment. This concern led to a series of discussions between the environmental community, industry, and government officials in an attempt to fashion a coordinated emission reduction strategy. Industry sought certainty to improve their planning processes and in exchange, the environmental community sought additional emission reductions of pollutants already regulated by the CAA in addition to regulation of Hg and CO_2.

The ongoing discussions influenced President Bush's campaign commitment. After he reneged, several pieces of legislation were introduced in the Congress to regulate the four pollutants. Recognizing he needed to do something to influence the debate, President Bush proposed the Clear Skies Initiative requiring reductions of SO_2, NOx, and Hg through cap-and-trade,[91] which was introduced as legislation in 2003. Competing legislation continued to be introduced requiring power plants to reduce emissions of the four pollutants. Neither approach passed and President

---

[90] Kossoy, A., G. Peszko, K. Oppermann, N. Prytz, N. Klein, K. Blok, L. Lam, L. Wong, B. Borkent. 2015. *State and Trends of Carbon Pricing 2015* (September), by World Bank, Washington, DC.

[91] Office of the Press Secretary, The White House. *Fact Sheet: President Bush Announces Clear Skies & Global Climate Change Initiatives.* (Washington, Executive Office of the President) 2002 February.

Bush attempted to implement his proposal to regulate the three pollutants through traditional CAAA regulation.

In hindsight, I assume that many power companies could have accepted four pollutant legislation and would prefer it to the regulations proposed by President Obama in climate change 2.0 to reduce CO_2 emissions from existing and new power plants. Companies generally support the certainty and flexibility offered by legislation in contrast to regulation, which is often inflexible and the product of an administrative process that is often more difficult to navigate. However, many would not take positions that conflicted with the Bush Administration and conservative industries opposition to CO_2 regulation.

Following the debate on multi-pollutant legislation, interest began to increase in expanding the cap-and-trade model from the power sector to the entire economy. Many in the power sector supported this believing they were singled out by the 4p approach and that covering more sources would create a larger market resulting in greater opportunities for cost-saving trade. The environmental community also supported expanding the model. Beginning in 2003, Senators John McCain (R-AR), Joseph Lieberman (D-CT), and several other Senators introduced GHG cap-and-trade legislation in the US. Similar legislation was reintroduced in subsequent Congresses. All received approximately 40 votes.[92]

In 2007, Senators Lieberman (I-CT) and Warner (R-VA) introduced the America's Climate Security Act, an economy-wide cap-and-trade program that would have covered 84% of US emissions and required reductions of GHG emissions 70% below 2005 levels in 2050.[93] The bill received a favorable vote in the senate Environment and Public Works Committee by an 11–8 margin, representing for the first time a congressional committee-approved climate change legislation. The legislation authorized the use of a significant volume of offsets for compliance and

---

[92] Senator John McCain (R-AZ), Senator Joseph Lieberman (D-CT) and several other US Senators introduced S. 139, *the Climate Stewardship Act*, beginning in 2003. It was defeated by a vote of 55–43. The legislation was reintroduced introduced in subsequent Congress's and received approximately 40 votes.

[93] Murray, B., M. Ross. *America's Climate Security Act: A Preliminary Assessment of Potential Economic Impacts*. Nicholas Institute for Environmental Policy Solutions, Duke University & RTI International. Report number: NI PB 07–04. 2007

incorporated greater detail regarding implementation, which had not been included in prior legislation.

In theory, expanding the trading program from the power sector, which was responsible for approximately 35% of national GHG emissions, to covering the GHG emissions from other industrial sectors such as manufacturing and transport, would increase the scope of reduction opportunities and help to lower costs of complying with emission reduction targets. As is illustrated in Chap. 5, theory and politics collided.

The inexorable march to cap-and-trade legislation was on. It was clear that the interest groups were preparing for a new Administration that would be less hostile to climate change than the Bush Administration. The legislative process emerging on climate change was similar to what I witnessed in the decade-long debate over acid rain legislation that was ultimately included in the CAAA of 1990. In this process, members of Congress introduce legislation, hearings are held to discuss it, greater understanding of its impacts is gained, new ideas emerge, and legislation ultimately passes. It seemed that climate change legislation might have been inevitable particularly since both Parties' nominees for the Presidency supported it.

Massachusetts Versus Environmental Protection Agency

Foreshadowing the future, it was not Congress that was the primary driver of US climate change policy. Ultimately, the judicial branch of government provided the foundation for the first large-scale federal effort to reduce national GHG emissions. On 2 April 2007, the Supreme Court made a decision in Massachusetts versus EPA that provided President Obama with the authority to regulate GHG emissions in the US.[94]

Twelve states and several environmental groups sued EPA over its refusal to regulate GHG emissions from new motor vehicles. EPA's refusal was in response to a 1999 petition to set standards for four GHGs

---

[94] Justice J. P. Stevens. *MASSACHUSETTS, ET AL., PETITIONERS v. ENVIRONMENTAL PROTECTION AGENCY ET AL.* Supreme Court of the United States. No. 05–1120. 2007.

emitted by new motor vehicles.[95] The DC Circuit Court ruled in favor of EPA on the issue. The case was then brought to the Supreme Court. Although there are many questions of law in such a far-reaching case, I will only note that the Supreme Court concluded that GHGs 'are air pollutants subject to regulation under the Clean Air Act'.[96] The EPA of the Bush Administration had argued it did not have the authority to regulate GHG emissions. In addition, the Supreme Court decided that a provision of the CAAA provides that EPA 'shall by regulation prescribe... standards applicable to the emission of any air pollutant from any class or classes of new motor vehicles or new motor vehicle engines, which in [the Administrator's] judgment cause, or contribute to, air pollution which may reasonably be anticipated to endanger public health or welfare'.[97] Taken further, if the EPA Administrator determines that an air pollutant causes or contributes to air pollution that endangers public health or welfare, the CAAA requires the agency to regulate the pollutant.

Chapter 6 describes how this decision influenced the new era of policy-making in the US. The Obama Administration rapidly determined that GHGs endangered public health and welfare.[98]

---

[95] Mendelson, J., A. Kimbrell. *Petition For Rulemaking And Collateral Relief Seeking The Regulation Of Greenhouse Gas Emissions From New Motor Vehicles Uunder ? 202 Of The Clean Air Act.* International Center for Technical Assessment. Available from: http://www.ciel.org/Publications/greenhouse_petition_EPA.pdf [Accessed 4 December 2015].

[96] Heinzerling, L. (2007), 'Massachusetts v. EPA'. *J. Envtl. Law and Litigation,* [Vol.22, 301 2007], 301–311.

[97] Justice J. P. Stevens. *Massachusetts v. Environmental Protection Agency.* Supreme Court of the United States. No. 05–1120. 2007.

[98] US Environmental Protection Agency. 40 CFR Chap. 1. *Endangerment and Cause or Contribute Findings for Greenhouse Gases under Section 202(a) of the Clean Air Act.* Available from: http://www3.epa.gov/climatechange/Downloads/endangerment/Federal_Register-EPA-HQ-OAR-2009-0171-Dec.15-09.pdf [Accessed 8 December 2015].

4

The End of Climate Change 1.0 Internationally

Overview

The rules and policies described in Chap. 3 led to the development of the global carbon markets. These included agreements reached at COP-7 in Marrakesh on the rules necessary to implement the CDM, the creation of the EU ETS, and the KP's entry into force. These steps spurred the initial market activity of climate change 1.0 by establishing market supply and demand. The end of Phase 1 of the EU ETS and the growing market for CERs in 2007 marked the close of this period. The burst of policy and market activity in Europe and at the international level was in stark contrast to the US scene at the federal level.

The period that followed began with great fanfare internationally with the onset of the KP period, Phase 2 of the ETS, and the promise of action in the US to reduce national GHG emissions for the first time. Its promise ended with the demise of the KP although some continued to implement it, turbulence in the EU ETS which required major reform to survive, and the defeat of cap-and-trade in the US. This period will ultimately be remembered for the end of climate change policy 1.0 and

© The Author(s) 2016   **91**
R.H. Rosenzweig, *Global Climate Change Policy and
Carbon Markets*, Energy, Climate and the Environment,
DOI 10.1057/978-1-137-56051-3_4

its top-down policy-model. Chapter 4 describes the turmoil that occurred in the EU market, the collapse of the CDM, and the conclusion of the KP commitment period.

This chapter begins with a review of the performance of Phase 2 of the EU ETS because it is a foundation of the EU's climate change strategy and the pillar of the global carbon markets. Two of the primary goals of a GHG cap-and-trade system are to achieve an environmental objective at the lowest cost while stimulating investment in low and non-emitting technologies necessary to decarbonize the economy during the twenty-first century. These can be competing goals. In order to accomplish them, prices for compliance instruments must remain in a band. They must be low enough to be politically acceptable and sufficient to avoid the lock-in of carbon-intensive technologies, which are often in operation for decades. And they must be high enough to provide certainty to incent long-term investment decisions in non- and low-emitting technologies while avoiding a political backlash. Striking this balance is difficult.

In all cases, the performance of environmental markets, including the EU ETS and its ability to achieve desired objectives, is impacted by the design features, interactions with other policies, and external dynamics. The other policies are frequently regulatory programs or mandates, some-times referred to as CPs. They operate alongside the market and can work at odds with each other. External dynamics can be rapid changes in eco-nomic conditions such as those that swept the globe during Phase 2 of the EU ETS. This chapter illustrates the challenges that EU confronted in designing a program capable of responding to these issues.

EU ETS: Preparing for Phase 2

Phase 1 of the EU ETS will be remembered for the precipitous price decline of EUAs in 2006 and their collapse to near zero in 2007.[1] Market participants expected Phase 2 to work better given their experience with the scheme, the resolution of design imperfections that created some of the problems in Phase 1, and other improvements that had been made.

---

[1] Ellerman, d., F. Convery, C. de Perthius (2010) Pricing Carbon: the European Union Emissions Trading Scheme, (New York and Cambridge, Cambridge University Press), PP. 140–142.

Importantly, the problem of inadequate emissions data (which the EC characterized as best guesses)[2] used to develop some member state's NAPs, which caused EUA prices to decline precipitously from €30 in April 2006 to less than €10 shortly thereafter, had been rectified.[3] The prohibition of banking surplus Phase 1 EUAs for use in Phase 2, which rendered them valueless, had also been addressed. Surplus Phase 2 EUAs could be banked for use in Phase 3.

Importantly, the EU and its member states had far more time to prepare for the implementation of Phase 2, including the development of NAPs, than in Phase 1. Because of the minimal amount of time between the directive establishing the EU ETS in October 2003 and its beginning in January 2005, the timing for EU member states to develop NAPs and gain approval by the EC was extremely tight.[4] The NAPs of Poland, the Czech Republic, Italy, and Greece, which accounted for nearly 30% of emissions covered by the EU ETS, had not been approved by the onset of the Phase 1 trading period.[5] The outcome of the NAP process in Phase 1 was to provide member states with 267 million more EUAs than verified emissions.[6] The system was long in Phase 1.

The process utilized in developing Phase 2 NAPs was much more orderly and the data was more precise.[7] It resulted in Phase 2 NAP totals approximately 12% below those included in Phase 1 NAPs and, importantly, approximately 5% below Phase 1 emissions.[8] Another important decision made during this period was to establish that regulated installations could use 1.4 billion CERs and ERUs to comply with their emissions limitation during the five-year period.[9]

---

[2] European Commission. *EU ETS 2005–2012. Phase one: 2005–2007.* Available from: http://ec.europa.eu/clima/policies/ets/pre2013/index_en.htm [Accessed 5 December 2015].

[3] Capoor, k., P. Ambrosi. *State and Trends of the Carbon Market 2007.* The World Bank and the International Emissions Trading Association, 2007. See figure 1 on P. 12.

[4] Ellerman, D. et al. (2010) *Pricing Carbon.* PP. 37–40.

[5] Ellerman, D. et al. (2010) *Pricing Carbon.* P. 39.

[6] Ellerman, D. et al. (2010) *Pricing Carbon.* P. 39. Table 3.4. P. 48.

[7] Ellerman, D. et al. (2010) *Pricing Carbon.* PP. 48–60.

[8] Ellerman, D. et al. (2010) *Pricing Carbon.* P. 54.

[9] Ellerman, D. et al. (2010) *Pricing Carbon.* For a description of the formula governing the use of CER/ERUs in Phase II see PP. 58 and 59.

In order to provide market participants with the certainty required to evaluate and make long-term investments, the EU also agreed in 2009 on the shape of Phase 3 of the system, which would run from 2013 to 2020.[10] Some of the important changes that were made included the creation of an EU-wide GHG emissions cap in lieu of national caps, a declining cap of 1.74% per year, a target for trading sectors of 21% below 2005 levels in 2020, and increasing the trading period from five to eight years. In addition, a much greater percentage of EUAs would be auctioned in Phase 3 in contrast to the free allocations that were generally provided in Phases 1 and 2.[11] However, other decisions increased uncertainty for EU ETS participants. These included conditioning a more stringent target on the outcome of the international negotiations to develop a successor to Kyoto and the volume of CERs and ERUs which would be allowed.[12]

Phase 2 Results

The following sections described what occurred in Phase 2 of the ETS.

Emissions

Emissions fell precipitously during Phase 2. The emissions of sectors covered by the EU ETS fell by more than 250 million tons from 2008 to 2012 or more than 10% during this period.[13] They were lower than allocated levels from 2009 to 2012.[14] These trends were prevalent in all key

---

[10] European Parliament and Council. 2009. DIRECTIVE 2009/29/EC of the European parliament and of the council of 23 April 2009 amending Directive 2003/87/EC so as to improve and extend the greenhouse gas emission allowance trading scheme of the Community. Brussels, European Parliament and Council.

[11] European Commission. *The EU Emissions Trading System*. Available from: http://ec.europa.eu/clima/publications/docs/factsheet_ets_en.pdf European Union, 2013.

[12] European Commission. *The European Union Emissions Trading System (EU ETS). Questions and Answers on the revised EU Emissions Trading System*. Available from: http://ec.europa.eu/clima/policies/ets/faq_en.htm [Accessed 26 December 2015].

[13] European Environment Agency. *EU Emissions Trading System (ETS) data viewer*. Available from: http://www.eea.europa.eu/data-and-maps/data/data-viewers/emissions-trading-viewer [Accessed 5 December 2015].

[14] European Environment Agency. *EU Emissions Trading System data*.

emitting sectors including combustion, iron and steel and cement production. The decline in emissions was greatest between 2008 and 2009;
the first full year of the global economic downturn.[15]

Prices

As would be expected in times of severe recession and falling emissions,
EUA prices declined throughout Phase 2. The reasons for this are discussed later. After prices reached €29 early in July 2008, they ended the
year below €16. They declined to approximately €8 in February 2009,
recovering to over €14 by the end of April. EUA prices were generally
stable for more than two years, ranging from €13 to € 16 from April
2009 until June 2011. Prices fell below €10 in November 2011, closed at
that level only four more times in 2011, and were most frequently in the
€6 to €9 range through the end of 2012.[16] Although this section focuses
on Phase 2, prices reached far lower levels in 2013 and 2014.

Trading Activity

Despite lower emissions, data indicates that traded EUA volumes increased
every year in Phase 2. They doubled from approximately 2.1 billion in
2007 to over 3.1 billion in 2008 and more than doubled again to over 6.3
billion in 2009. Traded volumes increased to over 7.9 billion in 2011.[17]
    It may seem counterintuitive that these increases in trade of EUAs
occurred as emissions and prices were falling. One explanation for this
provided by the WB suggested that industrials sold surplus allowances
to generate cash. Raising capital in the markets was difficult during this
period and credit markets were extremely tight.[18] Another explanation

---

[15] European Environment Agency. *EU Emissions Trading System data.*
[16] This data was generously shared with me by Thomson Reuters.
[17] European Commission. *EU ETS 2005–2012: Evolution of the European carbon market.* Available
from: http://ec.europa.eu/clima/policies/ets/pre2013/index_en.htm [Accessed 8 December 2015].
It is important to note that the EU provides different data of traded volumes in another document.
I used the higher volumes. The other data can be found at: http://ec.europa.eu/clima/publications/
docs/factsheet_ets_en.pdf
[18] Capoor, K., P. Ambrosi. *State and Trends of the Carbon Market 2009.* The World Bank. 2009. P. 6.

for increased trade provided by a former Natsource colleague was that volume increases when prices fall because traders are allocated a fixed budget for trading. Regardless of the reason, it seemed that increased trade was a win-win proposition for market participants. Allowance sales generated cash for industrials and, in theory, were profitable because they were allocated for free. Buyers were able to purchase EUAs for low prices and bank them for use to comply with more stringent caps in Phase 3.

A Growing EUA Surplus Demands Market Reforms

Toward the end of Phase 2, market participants knew that the EU ETS was in trouble and in need of significant reform. Supply and demand was out of balance. In its role of market participant, Natsource observed this imbalance first-hand. It was confirmed early in the trading period when many of our fund participants informed us that they had no further need for CERs. As a result, many of them opted out of the fund and some of our financial investors sought redemptions of their committed capital.

Confirming the markets supply demand imbalance, the EC presented a report on 14 November 2012 that concluded that the supply of compliance instruments (EUAs) and carbon credits (created by CDM and JI projects) exceeded emissions by more than 950 million tonnes at the end of 2011.[19] The report concluded that there could be as many as two billion surplus allowances in the system at the beginning of Phase 3.[20,21] As a result, prices would remain low for the foreseeable future. The paper concluded, '[b]ut with the surplus already at almost a billion allowances in 2011, there is a real risk of seriously undermining the orderly functioning of the carbon market by causing excessive price fluctuations due to the short-term over-supply of allowances'.[22] It is important to understand the

---

[19] European Commission. *REPORT FROM THE COMMISSION TO THE EUROPEAN PARLIAMENT AND THE COUNCIL The state of the European carbon market in 2012.* European Commission. COM(2012) 652 final, 2012. P. 4.

[20] European Commission. *The state of the European carbon market.* P. 5.

[21] According to a report developed by the European Environment Agency, *Trends and projections in Europe 2014 Tracking progress towards Europe's climate and energy targets for 2020,* the surplus did reach 2 billion in 2013. P. 28.

[22] European Commission. *The state of the European carbon market.* P. 6.

magnitude of this surplus. The EU allocated on average approximately two billion EUAs per year from 2008 to 2012.[23] The 2013 surplus could be equivalent to a year of allocated EUAs.

Two years later, a document developed by the EC indicated that the EUA surplus would grow to 2.5 billion by 2020.[24] In order to address this problem, the Commission had agreed to postpone or 'backload' the scheduled auctioning of 900 million EUAs in 2014, 2015, and 2016 until 2019 and 2020.[25] Although this would take supply off the market for a period of time, and reduce some of the surplus in the near-term, it would reintroduce the EUAs in 2019 and 2020. The EC recognized that although this could reduce volatility, it would not address the structural oversupply problem, which could only be addressed through longer-term reforms.[26] In order to address the longer-term supply/demand imbalances, a proposal was made to create a market stability reserve (MSR). The proposal was along the lines that follow: if it was determined that the EUA surplus in the market was below 400 million EUAs, 100 million EUAs would be released from the reserve and added to future auctioned volumes. In the opposite scenario, if volumes of EUAs in the marker were greater than 833 million, an amount equal to 12% of the surplus would be subtracted from auctioned amounts and added to the reserve unless the number of EUAs was less than 100 million.[27]

At the time of this writing, the EU Parliament and EU Council agreed to an MSR along the lines cited above with an important change. It would become operational in 2019 and endowed with the 900 million

---

[23] European Environment Agency. *EU Emissions Trading System data.*

[24] Commission Staff Working Document. *Impact Assessment Accompanying the Document, A policy framework for climate and energy in the period from 2020 up to 2030. Communication from the Commission to the European Parliament, the Council, the European Economic and Social Committee and the Committee of the Regions. A policy framework for climate and energy in the period from 2020 up to 2030* SWD (2014) 15 final: 2014. P. 25.

[25] European Commission. *Commission Regulation (EU) NO 176/2014 of 25 February 2014 amending Regulation (EU) 1031/2010 in particular to determine the volumes of greenhouse gas emissions allowances to be auctioned in 2013–2020.* Available from: http://eur-lex.europa.eu/legal-content/EN/TXT/?uri=uriserv:OJ.L_.2014.056.01.0011.01.ENG. [Accessed 8 December 2015].

[26] European Commission. *The state of the European carbon market.* PP. 6–7.

[27] For a brief description of the reserve, see International Emissions Trading Association, *The Market Stability Reserve: where are we with reform of the EU ETS.* 3 July 2015.

EUAs that were supposed to have been auctioned in 2019 and 2020.[28] In addition, any unallocated allowances in the 2013–2020 period would be added to the reserve. One account indicated that the market would be back in balance in the mid-2020s.[29]

The Causes of Surplus EUAs, EU ETS Reductions and Low Prices

In order to determine how to reform the EU ETS to ensure it functions better in the future, it is important to understand the variables which created the large surplus of EUAs that caused prices to decline to €6.47 at the end of Phase 2 and to even lower levels in the following years.

Before reviewing the analysis, it needs to be remembered that the EU ETS operates alongside other ambitious CPs including regulatory mandates. These include the Renewable Energy Directive[30] (RED) requiring the EU to meet 20% of its energy needs with renewables by 2020 and the Energy Efficiency Directive[31] (EED), which establishes a target of improving energy efficiency 20% by 2020. As described below, some believe that these programs had a significant impact on the performance of the EU ETS market.

Independent Analysis

According to independent analysis, GHG emissions were 1.1 or 1.2 GtCO$_2$, or 12%, lower at the end of Phase 2 than in the beginning.[32]

---

[28] Platts McGraw Hill Financial. *EU Council Adopts CO$_2$ Market Reserve to Start January 2019*. Available from: http://www.platts.com/latest-news/electric-power/london/eu-council-adopts-co2-market-reserve-to-start-26213345 [Accessed 8 December 2015].

[29] Platts McGraw *Hill Financial, EU Council Adopts CO$_2$ Market Reserve*.

[30] Directive 2009/28/EC of the European Parliament and of the Council of 23 April 2009 on the promotion of the use of energy from renewable sources and amending and subsequently repealing Directives 2001/77/EC and 2003/30/EC. Brussels, European Parliament and Council.

[31] Directive 2012/27/EU of the European Parliament and of the Council of 25 October 2012 on energy efficiency, amending Directives 2009/125/EC and 2010/30/EU and repealing Directives 2004/8/EC and 2006/32/EC. Brussels, European Parliament and Council.

[32] Gloaguen, O., E. Alberola. *Assessing the factors behind CO$_2$ emissions changes over Phase 1 and 2 of the EU ETS: an econometric analysis*, CDC Climate Research. Working Paper No. 2013–2015. 2013. P. 1.

This analysis attributed approximately 25% of the reductions to the recession, approximately 50% to the CPs, more than 40% to renewables, approximately 10–15% to improvements in energy efficiency.[33] Additional reductions were caused by fuel switching from coal to gas.[34] And in addition, installations regulated by the EU ETS used over 1 billion CERs and ERUs for compliance in Phase 2.[35]

Analysis by the same institution assumed in 2008 that sectors covered by the EU ETS would be required to achieve cumulative reductions in GHG emissions of five $GtCO_2$ from 2008 to 2020.[36] Combined, CPs were estimated to achieve 50% of the reductions required by the EU ETS sectors until 2020 with the policies implemented to achieve the RED accounting for two $GtCO_2$[37] of reductions and the EED estimated to achieve 0.5 $GtCO_2$ of reductions.[38] The use of 1.65 billion CERs and ERUs would achieve approximately 1.6 $GtCO_2$, or 32%,[39] of the estimated cumulative reductions. This left the ETS to achieve 18% of the reductions of GHG emissions prior to considering the impacts of the recession on baseline emissions. The analysis estimated that the recession lowered ETS emissions by one $GtCO_2$ or 20% of the required amount.[40] The conclusion from this analysis is that CPs, imports of CERs and ERUs, and the recession would achieve the five $GtCO_2$ necessary to achieve the EU ETS Phase 3 targets. The ETS was not required to achieve any reductions!

---

[33] Gloaguen, O., E. Alberola. *Assessing the factors behind CO_2 emissions changes over Phase 1 and 2 of the EU ETS.* P. 29.

[34] Gloaguen, O., E. Alberola. *Assessing the factors behind CO_2 emissions changes over Phase 1 and 2 of the EU ETS.* P. 29.

[35] European Environment Agency. *Trends and projections 2014, Tracking progress towards Europe's climate and energy targets for 2020.* EEA Report No 6/2014, 2014. P. 37.

[36] Berghhmans, N. *Energy Efficiency, renewable energy and CO_2 allowances in Europe: a need for coordination.* CDC Climate Research. Number 18. 2012. P. 1.

[37] Berghhmans, N. *Energy Efficiency, renewable energy and CO_2 allowances in Europe.* P. 4.

[38] Berghhmans, N. *Energy Efficiency, renewable energy and CO_2 allowances in Europe.* P. 6.

[39] Berghhmans, N. *Energy Efficiency, renewable energy and CO_2 allowances in Europe.* P. 6.

[40] Berghhmans. N. *Energy Efficiency, renewable energy and CO_2 allowances in Europe.* P. 6.

European Environment Agency (EEA) Analysis

I also reviewed analysis undertaken by the EEA that assessed the EU's progress in achieving its climate and energy objectives.[41] It indicated that emissions in the sectors covered by the EU ETS declined by approximately 19 % between 2005 and 2013, well on the way toward achieving the 21 % reduction target.[42] In general, it does not attribute these reductions to individual policies. It does conclude that the reduction in GHG emissions that occurred in 2013 from 2005 levels in the combustion sector resulted from increases in renewable energy.[43] The analysis also indicates that GHG emission reductions in the cement, iron, and refinery sectors resulted from reduced production due to the economic crisis, not from efficiency gains.[44] It concludes that the recession and project-based imports contributed to the low demand for allowances, the surplus, and low prices entering Phase 3.[45]

The independent analysis and EEA agree that imports of CERs and ERUs, and the economic downturn contributed to the EUA surplus and low allowance prices that dominated Phase 2 of the EU ETS and which continued in to Phase 3. However, in contrast, the independent analysis finds that the CPs played a much larger role in achieving reductions in Phase 2 and to the overall surplus and low prices. Separately, the European Commission attributes more of the EUA surplus to the economic downturn and imports of CERs/ERUs than to the EU ETS' interaction with other climate and energy policies, although it does acknowledge this to have played a role.[46]

---

[41] European Environment Agency. *Trends and projections 2014*.

[42] European Environment Agency. *Trends and projections 2014*. P. 28.

[43] European Environment Agency. *Trends and projections 2014*. P. 39.

[44] European Environment Agency. *Trends and projections 2014*. P. 39.

[45] European Environment Agency. *Trends and projections 2014*. P. 28.

[46] European Commission. *Communication from the commission to the European Parliament, The Council, The European Economic and Social Committee and The Committee of the Regions. A policy framework for climate and energy in the period from 2020 to 2030.* COM (2014) 15 final. 2014. P. 8.

The Interaction of Complementary Policies and the EU ETS

The European Environment Agency reviewed the status of each of the individual CPs in achieving their individual targets and undertook a qualitative assessment of some of the interactions between the EU ETS and the RED and EED.[47]

In describing some of the interaction between the CPs in a general fashion, the EEA focused on what it viewed as positive interactions between the policies. The analysis concludes, '[t]hese interactions should help improve policy design and implementation, correct for market failures and meet additional policy objectives'.[48] The EEA argued that the use of renewables resulted in 'avoided' GHG emissions, and less fossil fuel use while helping to provide other important economic, energy and environmental benefits.[49] Its analysis concluded that the low-carbon price in the EU ETS may have provided an incentive to member states to implement measures to increase the use of renewable energy sources.[50] I do not necessarily agree that it is positive that a low EUA price provided an incentive for member states to implement additional measures to benefit renewables, although market conditions may have dictated this.

Although renewables provide important co-benefits, the policies that induced their use can come at a high cost. Analysis was undertaken to determine the CO_2 abatement costs of renewable energy incentives provided to wind and solar in Germany. It found that wind costs to reduce CO_2 emissions averaged €44 per ton of CO_2 compared to EUA prices of approximately €11 from 2009 to 2010. The costs of abatement for solar averaged €537 per ton of CO_2 during this period, orders of magnitude higher than EUA prices.[51] At some point, the cost of CPs can cause a backlash and create public opposition to policies. The EEA goes on

[47] European Environment Agency. *Trends and projections 2014*. Section 7.

[48] European Environment Agency. *Trends and projections 2014*. P. 94.

[49] European Environment Agency. *Trends and projections 2014*. PP. 94–95.

[50] European Environment Agency. *Trends and projections 2014*. P. 94.

[51] Marcantonini, C., D. Ellerman. *The Cost of Abating CO₂ Emissions by Renewable Energy Incentives in Germany*. MIT Center for Energy and Environmental Policy Research. CEEPR WP 2013–005, 2013.

to state that energy efficiency 'may' also reduce demand for electricity in the EU ETS and the quantity of renewable energy required to meet targets for renewable energy sources.[52]

The analysis also reviewed the potential adverse impacts of the interaction between the policies. These included increasing uncertainty in the EU ETS because of the difficulty in predicting the success of the RED, and a reduction in the carbon price signal caused by an overachievement of the target in 2011–2012.[53] The energy efficiency targets could have the same impacts on the EU ETS as the renewable targets.

The analysis concluded by stating, '[e]stablishing an optimal mix and balance of policies and instruments, at national and European levels, requires optimisation of complementarities and minimisation of countervailing interactions between these policies'.[54] In order to achieve this objective, the EEA suggested the following: (i) identifying the effects of observed policies; (ii) analyzing the consistency in assumptions used by policy-makers to establish policies; and (iii) analysis of CPs interaction with EU ETS and non-EU ETS sectors.[55]

Combining market-based strategies and CPs such as those implemented to achieve the goals of the RED and EED, enables governments to pursue multiple objectives simultaneously. In theory, the approaches are complementary. Although markets can help to secure low cost reductions, CPs are required to address market barriers and failures that serve as obstacles to stimulating large-scale investments in renewables and improving energy efficiency and the multiple benefits they provide.[56]

The policy-model of combining market mechanisms and CPs to address climate change is increasing in prominence. Because of this, more knowledge is required regarding their interactions. Virtually all jurisdictions developing climate policies are likely to utilize this approach.

---

[52] European Environment Agency. *Trends and projections 2014*. P. 94.

[53] European Environment Agency. *Trends and projections 2014*. PP. 94–95.

[54] European Environment Agency. *Trends and projections 2014*. P. 96.

[55] European Environment Agency. *Trends and projections 2014*. P. 96.

[56] Rosenzweig, R., R. Youngman. *Exploring the Interaction Between California's Greenhouse Gas Emissions Cap-and-Trade Program and Complementary Emissions Reduction Policies*. Electric Power Research Institute. Report number: 3002000298, 2013. See P. 2–5 and Appendix B for arguments in favor of CPs.

As described in Chap. 3, the State of California developed a plan consisting of an economy wide cap-and-trade program and numerous CPs to achieve the State's GHG reduction target. It was estimated that the CPs would achieve the large majority of GHG reductions required to achieve California's GHG emissions reduction target.[57] In Europe, some analysis has reached similar conclusions.

In addition to the CPs potential impact in achieving large-scale GHG reductions in the sectors covered by the trading program and reducing the carbon price signal as occurred in the EU ETS, a few other important interactions are worth mentioning at this point. First, analysis has shown that achieving reductions in GHG emissions by CPs will likely come at a higher cost for society than a pure market-based system.[58,59] This is because mandates will reduce regulated firms' ability to take advantage of the flexibility inherent in the market. In addition, achieving GHG emissions reductions by improving energy efficiency and increasing the use of renewable energy through CPs could potentially hide the cost of climate change policies to the public. This is because similarly to what occurred in the EU ETS, these programs could lead to a reduction in demand for allowances/permits in a cap-and-trade program, distorting the price signal and lowering allowance prices. Whereas EUA prices are visible, the costs of CPs are not. So the public sees low allowance prices which do not tell the whole story of climate policies' cost.

As also seen by the EU ETS, if not implemented carefully, CPs may make it much more difficult for firms regulated by a cap-and-trade program to develop compliance and risk management strategies. This is because of the uncertainties as to whether CPs will achieve their reduction objectives. An example of this uncertainty, which was raised by the EEA analysis, is found within the power sector in the EU ETS.

---

[57] Two documents which illustrate the reductions that would be achieved by the CPs include California Air Resources Board. *Status of Plan Measures*. 2011 July and California Air Resources Board. *First Update to the Climate Change Scoping Plan, Building on the Framework Pursuant to AB 32, The Global Warming Solutions Act*. California Air Resources Board. 2014. PP. 92–93.

[58] See Section 5 of Rosenzweig, R., R. Youngman. *Exploring the Interaction Between California's Greenhouse Gas Emissions Cap-and-Trade Program and Complementary Emissions Reduction Policies* and Marcantonini, C., D. Ellerman. *The Cost of Abating CO_2 Emissions by Renewable Energy Incentives in Germany*, cited previously.

[59] This is highlighted previously.

If the renewable and efficiency directives achieved fewer reductions than forecast, emissions in the EU ETS would have been higher, increasing demand and therefore the price for EUAs. And, as observed in the EU ETS, the opposite can also occur. A confluence of events contributed to the surplus of EUAs that developed and lower demand. Although these uncertainties represent risks that business frequently manage, their derivation from policy uncertainties makes risk management more challenging.[60] This uncertainty also reduces incentives for firms to invest in the longer-term technologies required to address climate change.

Did the EU ETS Work?

The EU ETS has been in operation for a decade and is a foundation of the EU's climate policy, which has established the most far-reaching response to climate change in the world. The important question is: has the EU ETS worked? There are many criteria that can be used to answer this question, but I will focus on two important ones. First, to what extent did the EU ETS achieve reductions in GHG emissions within the sectors covered by the program? A review of academic analysis on this topic helps to address this question. Second, to what extent has the EU ETS incented long-term investment in low- and non-emitting technology? There is less literature on this topic.

Research Review of the Impact of the EU ETS in Reducing GHG Emissions

There is varied literature analyzing the extent to which the EU ETS reduced GHG emissions. A 2014 analysis[61] undertook a meta-review of articles seeking to answer this question. The first step in such analysis is

---

[60] This was alluded to in European Environment Agency. *Trends and projections 2014*. P. 94. See Section 6 of Rosenzweig, R., R. Youngman for a more thorough discussion of challenges in developing risk management strategies because of CPs and other issues in the context of California's climate change program.

[61] Laing, T., M. Sato, M. Grubb, and C. Comberti, Claudia (2014) 'The effects and side-effects of the EU emissions trading scheme,' *WIREs Clim Change* 2014, 5; 509–519. doi: 10. 1002/wcc. 283

the development of a counter-factual scenario of EU ETS sector emissions with and without the trading program. Eight studies, focused predominantly on the first four years of the ETS, created such scenarios for comparison.[62] Some of the studies were top-down and evaluated emissions without the EU ETS compared to verified emissions in the program. Others were bottom-up in nature, reviewing specific sector's BAU emissions compared to verified emissions. Those that used this approach also discussed the EU ETS with program participants.[63]

The authors of the meta-analysis indicate that the individual studies point to the EU ETS achieving reductions of 40–80 $MtCO_2$ per year, which is 2–4% of capped emissions.[64] This level of reductions is considered to be more than those achieved by other policy instruments including taxes, regulation, or voluntary initiatives.[65] Some of the studies concluded that the ETS contributed to a greater level of reductions.[66]

There is less research on EU ETS impacts following the financial crisis. The authors attributed this to the complexity of the economic crisis and the lag in the release of data.[67] The analysis concludes that the EU ETS may have stimulated reductions during the recession.[68] One study cited by the authors concludes that improvements in emissions intensity attributable to the EU ETS in 2008–2009 were 3.35% per year compared to 1% from 2006 to 2007, with greater improvements in power generation than industry.[69] It is unclear whether these analyses controlled for the impact of CPs.

A more recent study by a prominent analyst that has written extensively on the EU ETS and US emissions markets, attempts to evaluate the

---

[62] Laing, T., M. Sato, M. Grubb, and C. Comberti (2014) 'The effects and side-effects', See Table 1 on P. 511.

[63] Laing, T., M. Sato, M. Grubb, and C. Comberti (2014) 'The effects and side-effects', P. 511.

[64] Laing, T., M. Sato, M. Grubb, and C. Comberti (2014) 'The effects and side-effects', P. 511.

[65] Laing, T., M. Sato, M. Grubb, and C. Comberti (2014) 'The effects and side-effects', P. 511.

[66] See Table 1 for the results of all studies on P. 511 in T. Laing, M. Sato, M. Grubb, and C. Comberti (2014) 'The effects and side-effects'.

[67] Laing, T., M. Sato, M. Grubb, and C. Comberti (2014) 'The effects and side-effects', P. 511–512.

[68] Laing, T., M. Sato, M. Grubb, and C. Comberti (2014) 'The effects and side-effects', P. 512.

[69] Laing, T., M. Sato, M. Grubb, and C. Comberti (2014) 'The effects and side-effects', P. 512.

EU ETS performance from 2005 to 2012.[70] The counterfactual scenario can be assessed from a review of the analysis. The results conclude that emissions to GDP declined by 3.3% from 2004 to 2012 compared to 1% in the five years leading up to 2004.[71]

The study also analyzed EU ETS emissions from 2005 to 2012, including the additions of Romania and Bulgaria in 2007, Croatia in 2013, and Norway, Iceland, and Liechenstein at the beginning of Phase 2. It also accounted for increased emissions that resulted from adding installations and the aviation sector into the scheme. The analysis concluded that EU ETS emissions fell by 17.5% in 2012 from 2004 levels, the year prior to the beginning of the EU ETS although the result does not consider the impacts of CPs or improvements in energy efficiency that occurs on an annual basis.[72]

What does this mean? In general, the analysis concludes that the EU ETS contributed to emissions reductions particularly in Phase 1. The analysis cites fuel switching from coal to gas in the power sector as a cause of reductions but does not indicate others. There is less clarity regarding the EU ETS impacts in Phase 2; recent analysis suggests that reductions occurred, as did improvements in GHG emissions intensity. However, the study did not consider some key issues as indicated above.

Research Review of the Impact of the EU ETS in Stimulating Investment in Low and Non-emitting Technologies and Processes

This section is also based on the review of the meta-study of literature assessing the extent to which the EU ETS succeeded in stimulating investment in low and non-emitting technologies.[73] In short, the analysis

---

[70] Ellerman, D., C. Marcontonini, and Aleksandar Zaklan *The EU ETS: Eight Years and Counting.* Robert Schulman Centre for Advanced Studies, Robert Schuman Centre for Advance Studies, European University Institute, RSCAS2014/04, 2014.

[71] For a description of the methodology utilized for this study and results see P. 8. in D. Ellerman above.

[72] For a description of the methodology utilized for this study and results see PP. 8–9. in D. Ellerman above.

[73] Laing, T., M. Sato, M. Grubb, Michael and C. Comberti, (2014) 'Effects and side-effects'.

concludes that the EU ETS may have played a marginal role in focusing management's attention on the issue, and improving production processes. It also indicates that there was anecdotal evidence that the EU ETS helped to avoid investments in carbon-intensive technologies and increased research into longer-term technologies required to address climate change.[74] A survey of German firms concluded that although they had undertaken activities to reduce their GHG emissions, those actions were taken for other purposes.[75] Econometric analysis found that increased patenting, which often leads to innovation, was occurring more in firms regulated by the EU ETS than those that were not.[76]

The research indicated that although the EU ETS has had some beneficial technology impacts, it did not stimulate investment in the longer-term technologies that are necessary to address climate change. It also concluded that such investment was not likely in the future given the current price environment and the magnitude of surplus EUAs.

Given the price volatility of EUAs to date, and the low prices that have been prevalent for the past several years, it does not take much analysis to conclude that the system has not provided the certainty and price signals necessary to incent firms to make long-term investments in low- and non-emitting technologies. Higher prices of EUAs and greater certainty are required to achieve this objective.

Final Thoughts on the EU ETS

The EU ETS achieved some level of GHG emission reductions in its first two phases. However, its effectiveness in enabling the EU to achieve its long-term targets for GHG emissions will not be known for many years.

Based on my experience, the uneven performance of the EU ETS is not surprising. This can be attributed in part to the artificial nature of the environmental market, the lack of commercial and financial expertise of those that created it, specific design choices, interactions

---

[74] Laing, T., M. Sato, M. Grubb, and C. Comberti (2014) 'The effects and side-effects', P. 512.
[75] Laing, T., M. Sato, M. Grubb, and C. Comberti (2014) 'The effects and side-effects', P. 512.
[76] Laing, T., M. Sato, M. Grubb, and C. Comberti (2014) 'The effects and side-effects', P. 512.

with CPs, and the economic crisis. In my view, its largest single failure largest was caused by the EU's inability to respond to market conditions in a timely fashion. I mentioned previously that one account of the MSR indicated that the market would be in balance by the mid-2020s. If that is correct, it will have taken nearly 15 years to resolve the EUA supply overhang.

The EU ETS future is particularly important given that Phase 4 has been designed to achieve a GHG reduction target of 43% below 2005 levels from 2021 to 2030.[77] What follows are my views on the EU ETS to date and issues that require consideration moving forward.

The EU ETS Is a Political Success

Regardless of the challenges that the EU ETS experienced to date, it has been in operation for a decade at the time of this writing. This alone is a substantial achievement. It is still considered by the EU to be the cornerstone of its climate policy and it appears to have the support of those that it regulates. Although this may be primarily because most industrial firms prefer a carbon market to taxes, it does not matter. There is a far greater potential for success in mitigating climate change risks if the private sector is supportive of the policy prescriptions that are employed.

The Market Worked as It Should Have

The price of EUAs over the past several years has been too low to achieve significant reductions in GHG emissions from covered sectors and to provide incentives to stimulate large-scale investment in low- and non-emitting technologies. This is not an indictment of markets. Given the long running oversupply and resultant low EUA prices, the market has worked as intended.

---

[77] European Commission. *Revision for Phase IV (2020–2030)*. Available from: http://ec.europa.eu/clima/policies/ets/revision/index_en.htm [Accessed 8 December 2015].

Substantive Performance of the EU ETS Is a Function of Its Design, Interaction with Other Policies, and External Forces

The performance of Phase 1 was impacted by the rapid nature in which it was developed and its design. The performance of Phase 2 was impacted by its design, interactions with CPs, and the economic crisis. The performance of Phase 3 has been dictated to date by government's inability to respond in a timely fashion to the EUA surplus that has gotten worse over time.

It should not be surprising that environmental markets, including the EU ETS, act unexpectedly. Similar to all public policies, the ways in which they are designed impact their performance. Market mechanisms may be more prone to design problems given that policy-makers frequently do not have commercial experience.

Regarding Phase 4, the jury is out. There is no certainty that the MSR will work. One interesting attribute is that it recognizes government's inability to respond to market conditions. The volumes of EUAs that are auctioned will be adjusted automatically based on the volumes of supplies in the market.

There Needs to Be Better Understanding of the Interactions Between the EU ETS and CPs and Policy Coordination Moving Forward

The research cited in this section concluded that the policies put in place to achieve the objectives of the RED and EED have contributed to significant reductions in GHG emissions. However, these reductions have likely come at a higher cost than the EU ETS would have achieved, and contributed to the current EUA surplus, and perpetuated low prices. The CPs interaction with the EU ETS has the potential to create other adverse impacts as described above.

The following quotation summarizes the results in Phase 2 of the ETS that some believe were caused by the interactions between the CPs and the EU ETS: 'There is a risk that if the energy policies deliver too much of the abatement required to meet the [EU] ETS cap, the [EU] ETS allowance price could be reduced to the point where it no longer provides a clear signal for clean investment'.[78] This is exactly what occurred. In addition, uncertainties posed by the volume of reductions that CPs will achieve make it more difficult to estimate allowance prices, increasing the difficulty to evaluate investments in low and non-emitting technologies and to develop effective risk management strategies.[79]

Many supporters of the EU ETS have become increasingly concerned with the CP's impact on the functioning of the GHG emissions market. They are right to be concerned but must realize that governments have many objectives that compete with market efficiency. They will continue to put policies in place to achieve them.

It will be necessary to avoid what occurred during Phase 2 moving forward. And it will be challenging to do so because the EU has doubled down on these policies. The 2030 EU Climate and Energy Package includes targets of 27 per cent for renewable energy and improving energy efficiency 27% compared to projected levels of future energy consumption.[80] The policies put in place to achieve these goals will continue to exert significant influence on the operation of the EU ETS. The key to avoiding the pitfalls that have been experienced to date is for policy-makers to carefully consider the interactions between CPs and cap-and-trade systems, to design the CPs accordingly, and to review their performance regularly so that necessary adjustments can be made.[81]

Policy-Makers Need to Be Able to Respond to Changes in Market Conditions More Rapidly

Policy-makers and market participants knew about the growing imbalance in the EU ETS for many years. Yet it took several years longer to develop and implement proposals to reduce supply in the short-term and

---

[78] © OECD/IEA 2013, Managing Interactions between carbon pricing and existing energy policies, IEA Publishing. License: https://www.iea.org/t&c P. 21.

[79] © OECD/IEA 2013, Managing Interactions between carbon pricing and existing energy policies, IEA Publishing. License: https://www.iea.org/t&c P. 21.

[80] European Commission. *2030 Energy Strategy*. Available from: https://ec.europa.eu/energy/en/topics/energy-strategy/2030-energy-strategy [Accessed 8 December 2015].

[81] © OECD/IEA 2013, Managing Interactions between carbon pricing and existing energy policies, IEA Publishing. License: https://www.iea.org/t&c P. 21. Also see Berghmans, N., *Energy efficiency, renewable energy and CO₂ allowances in Europe and Reforming the EU ETS: give it some work!*, CDC Climate Research. Number 18: 2012 and Number 28: 2013.

attempt to normalize supply and demand in the long-term. During this time, the supply demand imbalance became much worse and its impacts became more widespread. Confidence in the market among its participants had to have declined significantly. If the prior estimate that the market will not be in balance until the mid-2020s is correct, 15 years will have been lost. This is an enormous amount of time.

One of the most detrimental shortcomings of ambitious top-down systems like the EU ETS is that decision-making is cumbersome and frequently slow. This makes it difficult to make necessary mid course corrections based on new knowledge and information. Yet, the ability to do so in a timely fashion is critical to successful public policy. This challenge is exacerbated by the structure of the EU and the cumbersome nature of its decision-making processes. The EU's inability to respond to market conditions in the EU ETS was a failure. It is unlikely that it could survive another prolonged crisis. In recognition of this, the MSR used an automatic mechanism based on the number of allowances in circulation at the time to determine whether the number of auctioned EUAs should be increased or decreased.

The slow reaction to addressing problems is not isolated to the EU ETS. This problem is common to systems which include elements of top-down systems.

Time for a Carbon Federal Reserve or Central Carbon Bank?

Central banks play a vital role in overseeing and in attempting to 'manage' the performance of national economies. The European Central Bank has some of the same responsibilities as a national central bank although there are differences with a central bank that oversees one economy. If the economy is underperforming, a central bank can reduce interest rates to stimulate demand and can use other tools to inject liquidity in to the economy to increase growth. When things get overheated and inflation becomes a concern, the central bank can raise interest rates to reduce demand in an attempt to cool things down. For example, the European Central Bank is currently engaged in significant efforts to revitalize the EU economy.

Carbon is another currency in the economy. Installations require EUAs to run their facilities. The inability to formulate a timely response as EUA prices declined significantly has adversely impacted the markets' performance.

However, there is no single regulator, such as a central bank, that has the authority to oversee the market and make necessary adjustments.

The concept of a carbon federal bank or some type of institution with authority to oversee the carbon market has been discussed at times and has been included in general forms in US legislation. Given the importance of carbon to the European economy, and the century-scale challenge that climate change presents, it makes sense to revisit the issue of whether such an institution could improve the markets' performance.[82] Providing authority to one institution to oversee the market could also go a long way to avoiding the recent delays in responding to the oversupply in the EU market and improving its future results.

The CDM: 2008–2012

The 2008–2012 period was key for the CDM, given it was in conjunction with the KP commitment period.

Market Details

As illustrated in Chap. 3, the CDM experienced rapid growth in 2005–2007. The market for contracted CERs increased to nearly 800 $MtCO_2e$ valued at almost $13 billion in 2007 with primary CERs accounting for approximately $7.5 billion or nearly 60 % of transacted value and the secondary market accounting for approximately $5.4 billion or over 40 % of market value.[83]

The CDM Declines from 2008 to 2012

Consistent with the trend in other markets during this period, the growth in the primary CER market stopped in 2008. Transacted

[82] de Perthius, C. *Carbon markets regulation: The case for a CO_2 Central Bank.* Climate Economic Chair CDC Climate and Paris Dauphine University. Number 10: 2011.

[83] Capoor, K., P. Ambrosi. *State and Trends 2009.* P. 31.

volumes of primary CERs declined by nearly 30 % in 2008 from 2007 levels to approximately 390 MtCO$_2$e and value declined by over 12 % to approximately $6.5 billion.[84] Yet at the same time, the market for secondary CERs exploded from 240 MtCO$_2$e, valued at over $5.45 billion in 2007, to nearly 1.1 billion MtCO$_2$e valued at over $26 billion in 2008, representing significant growth in both transacted volumes and value.[85] In general, growth was driven by compliance buyers purchasing secondary CERs at costs lower than for EUAs and financial firms purchasing for their own accounts with the goal of reselling at a profit.

After 2008, the bottom dropped out of the primary CER market. Transacted volumes for primary CERs fell to approximately 90 MtCO$_2$e in 2011, valued at approximately $1 billion,[86] both down over 80 % from their 2007 highs. Prices fell to approximately €1.50[87] cents in 2012 December and averaged €0.37 cents or 0.51 cents in 2013.[88] The CDM was in free-fall from which it would not recover.

What Happened to CER Demand

The confluence between increased supply of compliance instruments and falling demand led to the deteriorating condition of the CDM. The enormous EUA surplus at the end of 2012 was forecast to increase to over 2.5 billion in 2020.[89] At the same time the EUA surplus was growing, supplies of CERs and ERUs increased dramatically. CER issuances increased to over 300 million per year in both 2011 and 2012, by far the highest

---

[84] Capoor, K., P. Ambrosi. *State and Trends 2009.* P. 31.

[85] Capoor, K., P. Ambrosi. *State and Trends 2009.* P. 31.

[86] Kossoy, A., P. Guignon. *State and Trends of the Carbon Market 2012.* The World Bank 2012. P. 49.

[87] Turner, G. *Carbon Market Update.* [Presentation] to the 12th Annual IEA-IETA-EPRI Annual Workshop on Greenhouse Gas Emissions Trading. Paris. 15 October 2012.

[88] World Bank. 2014.State and Trends of Carbon Pricing 2014. Washington, DC: World Bank. doi:10.1596/978-1-4648-0268-3. P. 39.

[89] Commission Staff Working Document. *Impact Assessment Accompanying the Document, A policy framework for climate and energy in the period from 2020 up to 2030.* P. 25.

levels to date.[90] And over 500 million ERUs were issued in 2012, which was far higher than the combined amount in all prior years.[91]

At the same time that supplies increased, there was no demand for CERs in the EU ETS. There was no economic benefit to take the risk to purchase primary CERs once EUAs prices plunged in 2011. Demand was also lower in Japan. Things deteriorated even further after 2012.

Market Reality Hits Home

At Natsource, we did not need to analyze the data to understand what was occurring in the market. We experienced it first-hand. For the most part, participants in Natsource's largest fund, primarily industrial firms from Europe and Japan, did not want any more CERs, even those that had been contracted. I had previously left as COO in 2011 after helping to downsize the company. With many of our staff gone, I returned to assist during the period of market turmoil. We fielded calls regularly asking if there was a way we could avoid taking delivery of CERs from contracted projects. If we determined there was a material breach, we terminated the purchase contract. This was occurring all throughout the market.

Two of our most prominent transactions, described in Chap. 3, felt the effects of this market turmoil. Participants in the World Bank UCF came to an agreement with the two project owners to terminate the ERPAs in 2013, a year earlier than if they had run their course.[92] The members of its syndicate, which included Natsource entities, welcomed this settlement. In addition to participants in our funds not needing the CERs, the EU prohibited the use of those CERs created by HFC-23 and N20 projects in 2011, unless they were created by existing projects by the end of 2012 and used to comply with 2012 ETS targets by 30 April 2013.[93] The later year vintage would have no value.

---

[90] World Bank. 2014. *State and Trends of Carbon Pricing 2014*. Figure 6. P. 39.

[91] World Bank. 2014. *State and Trends of Carbon Pricing 2014*. Figure 6. P. 39.

[92] The World Bank. *Statement September 12, 2013*. Available from: https://wbcarbonfinance.org/ Router.cfm?Page=UCF [Accessed 9 December 2015].

[93] European Commission. *Emissions trading: Commission welcomes vote to ban certain industrial gas credits*. [Press Release] 21 January 2011.

I had previously described the call option provision in the Shenma ERPAs that provided our financial investors an option not to purchase a portion of the projects CER volumes. When market prices collapsed, the financial buyers declined to buy any of the project's CERs, with the predictable result that the project entity elected to sell them to the compliance buyers.

As is typical in large firms, many of the participants' staff that were responsible for GG-CAP when the Shenma transaction was completed in 2006 had either been reassigned to other activities or were no longer with their company. When it became clear that the compliance buyers would be required to purchase the additional volumes of CERs, many argued that they could not be required to purchase what they did not need. This occurred although the GG-CAP Investment Committee, which included two members appointed by the fund participants including a representative from the company that made the largest purchase commitment, had in 2006 unanimously approved the deal and fund participants voted to support it. Natsource also had to purchase a larger volume because of the incentive compensation provision described earlier. This was an extremely challenging issue to resolve.

The Impact on Natsource and the Industry

Firms lose money when markets crash. The infrastructure of consultants, designated operational entities, brokers, carbon specialist firms like Natsource, investment banks, and others that were created to serve the carbon market since 2000 were battered. There were consolidations, bankruptcies, and significant layoffs within these firms and those that owned regulated installations.

This period was the beginning of the end for Natsource. We were hit with a wave of redemptions from financial investors that resulted in reduced fees. In addition, participants in GG-CAP, and NAT-CAP, a smaller fund, exercised their rights to opt out of further purchase commitments because they had no need for CERs. The redemptions and opt outs caused a loss of management fees that in theory were supposed to cover the costs of fund management. Even though participants left the funds, Natsource retained its obligations to manage purchase contracts we had entered into on the funds' behalf, perform settlement of deals, and report performance. Given that GG-CAP was a first-of-its-kind mechanism, we had not fully contemplated these events. Because of this, we did not negotiate exit fees from those that opted out in the initial contract, although we retained a number of obligations and these firms expected continued services. We were able to negotiate small maintenance fees with participants that opted out.

Another loss resulted from the price risk Natsource was exposed to as CER prices fell. In Chap. 3, I described that Natsource's incentive compensation for managing GG-CAP was to purchase a fixed percentage of delivered CERs from participants at their acquisition cost. We would then sell the CERs in the market at a profit. In the perfect world of continually increasing prices, this is a profitable approach. And we were able to capitalize on price appreciation for a few years. However, we were exposed to falling market prices. The fund had contracted for over 25 million CERs by the time the market started to decline. In a hypothetical 4% purchase obligation, we would be required to purchase one million CERs. Assuming a purchase price of €7, and a sales price of €2, we would lose €5 million. And CER prices fell to levels lower than that. The combination of lower revenues, increased expenditures to purchase CERs, and the need to maintain staff to implement contractual obligations impacted the company.

We could have weathered these challenges. However, the larger problem was that there were no opportunities to make money because of the lack of demand for CERs/ERUs and little prospect for market recovery. Rather than suffer greater losses and deplete our cash reserves, the company's management decided to close Natsource during 2012. It made more sense to distribute remaining cash to the company's investors than to spend it. And management took pride that we were able to do so.

Natsource officially closed on 31 December 2014. The last two years were spent unwinding projects, closing offices, dealing with threats of litigation, and closing funds and associated companies.

One of the continuing effects of the market decline was the dissolution of the infrastructure that was created to support the market.

In the pre-2008 period, growth occurred as firms attempted to capitalize on market growth. Companies went public, investment banks like JP Morgan Chase and Socgen acquired CDM development companies; others developed significant origination capabilities and traded for their own accounts. Consultancies supported these efforts. By 2012, the carbon specialist companies had ceased most market activities, diversified in to other markets, or had been taken over. Investment banks sold off the companies they acquired and ceased their trading operations. DOEs and consultancies closed. The point of this is that these firms were critical to the markets growth and functioning. It will be difficult and more costly to recreate a similar infrastructure in the future that is necessary to support the market.

The CDM by the Numbers

The CDM was a source of significant controversy. It is important to step back and view the results of the mechanism from its inception through 2012. This review will stop at 2012 because the analysis of its performance is through such year and its continued deterioration in the years following 2012 is well documented.

Registered Projects and Project Types

By the end of 2012, 7165 projects were registered by the CDM.[94] Over 50% were located in China, and nearly 20% were located in India.[95]

Over 70% of registered projects were renewable energy and surprisingly, less than 2% were industrial gas.[96]

---

[94] CDM Insights. *Project Activities*. Available from: http://cdm.unfccc.int/Statistics/Public/ CDMinsights/index.html [Accessed 9 December 2015].

[95] United Nations Framework Convention on Climate Change. *Project Activities*. Available from: http://cdm.unfccc.int/Statistics/Public/CDMinsights/index.html [Accessed 9 December 2015].

[96] UNEP DTU Partnership. *CDM/JI Pipeline Analysis*. Available from: http://www.cdmpipeline. org/[Accessed 9 December 2015].

Issued CERs and Asset Classes

The registered projects resulted in 1.155 billion CERs,[97,98] through 2012. Approximately 61% were created in China and 14% were created in India.[99]

Industrial gas projects created over 60% of issued CERs through 2012, approximately 40% by HFCs and approximately 20% by N2O.[100] Eleven percent were created by hydro projects, wind projects were responsible for 9%, and methane created 7%.[101]

Investment

It is estimated that $28 billion in pre-2013 CER contracts supported more than $130 billion in investment.[102] This amount is greater than the total of all overseas development assistance in 2011.[103]

The CDM: Did It Work?

The two primary criteria for evaluating the success of the CDM come from the objectives included in Article 12 2. of the KP. Namely, did the CDM assist developing countries achieve sustainable development and assist Annex I countries comply with their emission reduction targets?

---

[97] United Nations Framework Convention on Climate Change. *Project Activities*. Available from: http://cdm.unfccc.int/Statistics/Public/CDMinsights/index.html [Accessed 9 December 2015].

[98] Two other prominent databases, IGES and CDM pipeline, show 1.5 billion issued CERs through 2012. However, I use the data provided by CDM that is published on the UNFCCC website.

[99] United Nations Framework Convention on Climate Change. *Project Activities*. Available from: http://cdm.unfccc.int/Statistics/Public/CDMinsights/index.html [Accessed 9 December 2015].

[100] United Nations Framework Convention on Climate Change. *Project Activities*. Available from: http://cdm.unfccc.int/Statistics/Public/CDMinsights/index.html [Accessed 9 December 2015].

[101] United Nations Framework Convention on Climate Change. *Project Activities*. Available from: http://cdm.unfccc.int/Statistics/Public/CDMinsights/index.html [Accessed 9 December 2015].

[102] Kossoy, A., P. Guignon. *State and Trends 2012*. P. 49.

[103] World Bank. 2014. *State and Trends of Carbon Pricing 2014*. P. 45.

Sustainable Development

Based on my experience, I share the view held by many sustainable development (SD) advocates that the CDM failed in achieving this objective.[104]

In the negotiations over the CDM, developing countries argued successfully, appropriately in my view, that they should have the authority to determine whether a CDM project met its SD criteria. There is not a one-size-fits all approach to this question. Each country had different SD objectives that were unique to their own circumstances and objectives they were attempting to achieve. As a result, there was no international standard to determine whether CDM projects conformed to host country SD criteria prior to securing an LOA. The matter was left to each host country to decide. After the LOA was issued, the DOE would then review the PDD and the SD benefits in the validation phase. If the project activity was registered by the CDM, there was no ongoing process to measure and verify whether the SD benefits were achieved. This is distinct from the project implementation phase which required that the project proponent adhere to a monitoring plan included in the PDD and verification be made by a DOE that the project activity was reducing GHG emissions.

A consistent theme in the literature on this topic is that because of the competition to attract CDM projects among countries, DNAs set the SD bar low and did not raise it in the project evaluation phase. Some analysts referred to this as the race to the bottom.[105] In their evaluation of a CDM project, DNAs used a qualitative checklist to determine if a project met the DNAs SD criteria.[106]

Quantitative analyses have been undertaken in an attempt to measure the CDM's SD contributions. I will briefly review the results of two such analyses without describing the specific methodologies that

---

[104] Olsen, K., H. *The Clean Development Mechanisms Contribution to Sustainable Development: A Review of the Literature.* UNEP Riso Centre: Energy Climate and Sustainable Development, Riso National Laboratory. 2005 [?].

[105] Kelly, C., N. Helme. Ensuring *CDM Project Compatibility with Sustainable Development Goals.* Working paper. Center for Clean Air Policy, Washington, 2000.

[106] Olsen, K., H, J Fenham. *Sustainable Development Benefits of Clean Development Projects.* UNEP Riso Centre, Energy Climate and Sustainable Development, 2006. P. 3.

were employed. The first assessed whether 16 projects registered as of 30 August 2005 met the CDM's twin objectives, achieving GHG reductions and contributing to developing countries SD.[107] The SD evaluation was based on the projects' contributions to employment generation, equal distribution of CDM returns, and improvement of local air quality.[108]

The analysis concluded that 72 % of the CERs in the portfolio received an A ranking for additionality.[109] This would be expected because two large projects in the portfolio were HFC reduction, which was considered by most, even critics, to be additional. Three landfill gas projects also scored well in the additionality rating.[110] These projects generally scored lowest in the SD categories.[111]

The 11 remaining projects had a low probability of being considered additional.[112] With respect to the three specific measures of SD, the study concluded that 11 of the 16 projects scored low on employment generation, the large majority of projects received a good rating for their distribution of CER revenues, and 11 of the non-additional 16 projects received an A or B for improving local air quality.[113]

The analysis concluded that the market assigned a price to GHG reductions, but not to sustainable development and that none of the projects contributed strongly to achieving both of the CDMs objectives.[114]

The second analysis assessed the additionality and SD benefits of 40 registered projects in India as of 1 January 2009, while also attempting to

---

[107] Sutter, C., J. C. Perreno (2007) 'Does the current Clean Development Mechanism (CDM) deliver its sustainable development claim? An analysis of officially registered CDM projects'. *Climatic Change*, Volume 84, Issue 1, 75–90.

[108] Sutter, C., J. C. Perreno (2007) 'Does the current Clean Development Mechanism (CDM) deliver its sustainable development claim?', PP. 79–80.

[109] Sutter, C., J. C. Perreno (2007) 'Does the current Clean Development Mechanism (CDM) deliver its sustainable development claim?', P. 85.

[110] Sutter, C., J. C. Perreno (2007) 'Does the current Clean Development Mechanism (CDM) deliver its sustainable development claim?', P. 86.

[111] Sutter, C., J. C. Perreno (2007) 'Does the current Clean Development Mechanism (CDM) deliver its sustainable development claim?', P. 87.

[112] Sutter, C., J. C. Perreno (2007) 'Does the current Clean Development Mechanism (CDM) deliver its sustainable development claim?', P. 86.

[113] Sutter, C., J. C. Perreno (2007) 'Does the current Clean Development Mechanism (CDM) deliver its sustainable development claim?', P. 87.

[114] Sutter, C., J. C. Perreno (2007) 'Does the current Clean Development Mechanism (CDM) deliver its sustainable development claim?', P. 89.

determine if there is a tradeoff between the CDMs two objectives.[115] The 31 small-scale and 9 large-scale projects evaluated included: 15 biomass, 12 wind, 7 hydro, 4 energy efficiency, and 2 HFC-23.[116] Additionality was analyzed by assessing the impacts of CERs on projects' internal rate of return and SD was measured by using 11 criteria included in the categories of social, environmental, and economic development.[117]

The analysis determined that hydro and wind scored the lowest in the additionality category while biomass scored a little higher. It was determined that these projects provided a high level of SD benefits. HFC and energy efficiency projects were rated the exact opposite. They scored the highest in additionality but provided the lowest level of SD benefits.[118]

Most importantly, and similar to the prior analysis, no projects were found both to be additional and to contribute to SD.[119] Twenty-seven of the projects or nearly 70% were determined to have high levels of SD benefits but were not determined to be additional on a relative basis, four projects were likely to be additional and provide minimal SD benefits while the remaining nine projects scored low in both categories.[120]

The results of the two quantitative analyses are similar. Although the analyses are limited by sample size and methodological issues, they are also consistent with critics' views that the CDM has not been successful in contributing significantly to the host countries SD, particularly in early times, and that there appears to be a tradeoff between achieving low cost GHG reductions and SD. It is possible that the CDM may have contributed more to SD once industrial gas projects were exhausted

---

[115] Alexeew, J., L. Bergset, K. Meyer, J. Petersen, L. Schneider and C. Unger (2010) 'An analysis of the relationship between the additionality of CDM projects and their contribution to sustainable development', *International Environmental Agreements: Politics Law and Economics*, Volume 10, Issue 3, 233–248.

[116] Alexeew, J., L. Bergset, K. Meyer, J. Petersen, L. Schneider and C. Unger (2010) 'An analysis of the relationship', P. 235.

[117] Alexeew, J., L. Bergset, K. Meyer, J. Petersen, L. Schneider and C. Unger (2010) 'An analysis of the relationship', PP. 236–241.

[118] Alexeew, J., L. Bergset, K. Meyer, J. Petersen, L. Schneider and C. Unger (2010) 'An analysis of the relationship', PP. 241–243.

[119] Alexeew, J., L. Bergset, K. Meyer, J. Petersen, L. Schneider and C. Unger (2010) 'An analysis of the relationship', 243–244.

[120] Alexeew, J., L. Bergset, K. Meyer, J. Petersen, L. Schneider and C. Unger (2010) 'An analysis of the relationship', PP. 243–244.

early in the KP commitment period and renewable and efficiency projects increased in prominence. Based on the analysis of SD that was cited, the issue would have been whether these projects were also additional.

Because of the dissatisfaction with the results of the checklist used by DNAs and challenges with others, experts proposed alternative approaches to improve the CDM's ability to contribute to the host country's SD. One such approach was to utilize an international standard that would include a 'taxonomy' to determine CDM SD benefits.[121] The framework included the categories of environmental, social, economic, and other benefits with specific criteria and a mix of qualitative and quantitative analysis. To test the framework, a study was conducted to determine the number and type of SD benefits of CDM projects in the pipeline as of 3 May 2006.[122] The approach ultimately inspired the CDM to develop an SD tool.[123] However, the claim of SD benefits in the PDD were not monitored or verified and stakeholders are not provided the opportunity to describe benefits.[124]

The SD issue was controversial. Although contributing to developing country's SD was one of the CDMs' two objectives, it has been established in the literature that the creation of cost effective CER supply took precedence over SD from its inception. The literature is correct. The single objective of our fund participants was to comply with its GHG emission reduction requirements at as low a cost as possible. Our responsibility as Fund Manager was to secure the lowest cost CERs on their behalf. There was a provision in one of our management contracts that enabled a participant to 'to provide a notice of objection regarding any ERPA [Emission Reduction Purchase Agreement] that conflicts with its then current social, ethical or other policies'. A notice of objection would have prohibited the fund from entering in to any transaction in which

---

[121] Olsen, K. H., and J. Fenhann (2008) 'Sustainable development benefits of CDM projects: A new methodology for sustainability assessment based on text analysis of the project design documents submitted for validation', *Energy Policy*, 36, 2819–2830.

[122] For a description of the study, its methodology and results see Olsen K. H., and J. Fenhann, PP. 2823–2830.

[123] United Nations Framework Convention on Climate Change. *Voluntary tool for describing sustainable development co-benefits (SDC) of CDM project activities or programmes of activities (SD Tool)*. Available at: http://cdm.unfccc.int/Reference/tools/index.html. [Accessed 9 December 2015].

[124] Olsen, K., H. *CDM sustainable development co-benefit indicators*. [presentation] UNEP. 4–6 December 2012.

such a notice was filed. Although our funds entered into many contracts, no participant ever filed such an objection because of a conflict with its current social or ethical policies.

The CDM's emphasis on low cost supply was entirely predictable for various reasons. First, private firms and governments with emission reduction targets were primarily concerned with controlling costs. This concern was amplified given the lack of US participation in the KP and developing country commitments to reduce their GHG emissions. Any effort to control GHG emissions that caused economic harm and/or benefited competitors that did not confront similar regulation would not be politically sustainable. In addition, increased costs resulting from a GHG control program are easy to claim and can usually be quantified in some fashion. In contrast, and as illustrated in this section, SD remains an elusive concept. It means many things to different people. It is difficult to define, operationalize, and quantify.

Assisting Annex I Parties Comply with Their Emission Reduction Obligations

The other objective of the CDM was to assist developed countries achieve their emission reduction requirements. The consensus is that the CDM lowered developed countries' costs to comply with their emission reduction requirements.

One analysis was undertaken by the UNFCCC.[125] It concluded that the CDM reduced compliance costs of EU ETS installations, EU governments, and Japanese industry by approximately €2.850 billion ($US 3.60) from 2008 to 2012.[126] For several reasons explained in the analysis, the estimates of cost savings were likely conservative. One reason cited was that the CDM helped to reduce costs more than assumed in the EU ETS because CER supplies likely contributed to lower EUA prices.[127]

---

[125] Kirkman, G., S. Seres, E. Haites, and R. Spalding-Fecher. *Benefits of the Clean Development Mechanism 2012.* United Nations Framework Convention on Climate Change. 2012.

[126] Kirkman, G., S. Seres, E. Haites, and R. Spalding-Fecher. *Benefits of the Clean Development Mechanism 2012.* The cost savings and the methodologies used to calculate them can be found on pages 55–59.

[127] Kirkman, G., S. Seres, E. Haites, and R. Spalding-Fecher. *Benefits of the Clean Development Mechanism 2012.* P. 57.

Additional analysis concluded that CER and ERU purchases reduced EU ETS compliance costs by approximately €4 billion from 2008 to 2012.[128] The CER savings of approximately €2.4 billion assumed by the analysis is low.[129] The analysis also cites another model which estimates that the CDM and JI contributed to cost savings for EU ETS installations of nearly €21 billion during the period 2008–2012 because EUA prices would have been higher if CERs and ERUs did not exist.[130]

Based on these two analyses and Natsource's experience in the market, it is reasonable to conclude that the CDM lowered the costs of complying with GHG emission reduction targets.

The CDM Succeeded, The CDM Failed

Chapters 3 and 4 reviewed the growth and demise of the CDM market from 2005 to 2012, and whether the mechanism achieved its objectives. It also illustrated how a private firm like Natsource, which executed contracts to purchase CERs valued in excess of $1billion, participated in the market and was affected by its downfall.

I want to provide my views on the successes and failures of the CDM. Some of the lessons derived from the CDM are macro in nature and will be generally applicable to other offset systems, while others are unique to its implementation. The macro related issues are presented first. It is important to learn from the CDM in order to realistically assess the potential of GHG offset systems in climate change 2.0 and in developing the rules to govern the new mechanism included in the Paris Agreement.[131]

---

[128] Stephan, N., V. Bellassen, and E. Alberola. *Use of Industrial Credits By European Industrial Installations: From An Efficient Market To A Burst Bubble.* CDC Climate Research. No. 43. 2013. P. 19.

[129] This is because the savings were calculated by multiplying the amount of surrendered CERs during the five-year period by the spread in EUA and secondary CER prices. Many CER buyers purchased large amounts of primary CERs which cost less than secondaries. The spread between EUA prices and primaries was larger resulting in greater savings.

[130] Stephan, N., V. Bellassen, and E. Alberola. *Use of Industrial Credits By European Industrial Installations.* PP. 19–20.

[131] *The Paris Agreement to the United Nations Framework Convention on Climate Change.* Article 6. United Nations. 2015.

The CDM Was Unable to Respond to Problems in a Timely Fashion

The ability to implement successful public policies requires the ability to learn, adapt to new knowledge and information, and to make midcourse corrections. This is particularly relevant to climate change given continued improvements in the understanding of its causes, and its temporal dimension. The CDM was an ambitious learning by doing mechanism. It is impressive that it was developed and operationalized in less than a decade even though there had not been any experience with similar mechanisms. Although much was learned from its operation, adequate processes were not in place to solve systemic problems in a timely fashion.

For example, problems with the project cycle including the application of additionality, and overall decision-making were identified as concerns early in the CDM's operation. Solutions could have been devised and implemented earlier had there been better processes in place to do so. This could have increased stakeholder confidence in the CDM, improved its performance and political viability during the KP. Instead, its failure to do so was a leading cause of its demise and the perception that global mechanisms may not be able to work.

The failures of the CDM to adapt to new information and implement necessary changes in a timely fashion are not unique. They are common to top-down, complex systems. As described earlier in this chapter, the EU ETS suffered from the same problem in failing to address the large surplus of EUAs with significant consequences. However, the UN's administration of the mechanism, combined with the cumbersome, inflexible decision-making that characterized the top-down international climate change negotiations exacerbated the CDM's challenges. Some of the KP's and CDM's flaws—and those that plagued climate change 1.0 in general—can be attributed to their top-down characteristics, inefficient decision-making that required achieving consensus among dozens of countries, and not having mechanisms in place to learn and adapt.

The CDM Was the Subject of Extreme Controversy: Any Large-Scale Offset System Will Be

The CDM was controversial since its inception. Critics argued that the system was too lax while developers vented their frustration with the project cycle and many aspects of the mechanisms' administration.

As will be required by any offset system, the creation of a counter-factual scenario to establish BAU emissions, demonstrating that emission reductions were achieved from BAU, and that they were additional to what would have occurred only because of a project will almost always be inefficient, controversial, and subject to criticism. There is no way to prove with 100 % assurance that a project reduced emissions from what they would have been under business-as-usual. It is true in the CDM and would be in any other project-based mechanism, regardless of whether it is at the global, national, or sub-national level. And as it regards additionality, it is unrealistic to expect that projects will only be implemented because of their ability to create offsets.

Similar to satisfying concerns with environmental integrity, policy-makers will be challenged to design a project cycle that satisfies developers' needs for efficient and timely decision-making. And because the revenues generated by offset sales are often pivotal for project implementation, this is an important issue that is often overlooked. The lengthier the process, the more likely the developer will lose interest. Approaches that attempt to strike the balance between environmental integrity while expediting decision-making, such as positive lists, standardization and others can help. But in general, the commonly held view is that rules will either be so tight that few projects will be approved or so loose that too many non-additional projects to be approved.

One of the requirements in developing an effective and sustainable program is to strike the delicate compromises necessary to gain and maintain the support and acceptance of participating stakeholders. This enables it to weather attacks on its credibility. The CDM was never able to do so and in this regard, it must be viewed as a failure. These challenges are significant in any realm. They are exacerbated in a global mechanism like the CDM by the lack of leadership, the daunting challenge of achieving consensus amongst more than a 100 countries, and, most importantly, the inability of top-down systems to respond to new knowledge and information in a timely fashion.

The controversies surrounding the CDM are not a recommendation for policy-makers to avoid offset programs. They can help to control costs and stimulate investment and innovation in sectors not covered by a trading program. However, the mechanisms will often be controversial, inefficient, resource intensive, and unlikely to provide all of the economic benefits that analysis usually forecasts. It is the nature of the beast. This is particularly the case with offset systems created to achieve reductions at the project level.

Conditions Necessary for Private Sector Engagement

Supporters of the CDM point to the positive data cited above as evidence of the CDM's success, most notably: the value of primary CER contracts, the billions of dollars leveraged in project investment, and the creation of over 1.150 billion CERs. These are significant accomplishments. In addition, some point to the private sectors' participation in the CDM as a key measure of its success and evidence that it would enthusiastically participate in a revised CDM or follow-on mechanism at the global level.

The early days of the CDM was an exciting time for those that believed that market mechanisms could be part of the solution to successfully mitigate climate change. Entrepreneurial firms like Natsource, technology companies, and the banking and financial community, including hedge funds and private equity, were all engaged in the carbon market. And they needed to be. Massive amounts of private capital and innovation are required to achieve the large-scale mitigation required during the twenty-first century to achieve the GHG emission reductions required to limit temperature increases. Only the private sector, including the financial community, can supply the resources, the entrepreneurial spirit, and the innovation that is required to make a dent in the problem. There is not enough public money to do so.

The private sector did not show up because of the CDM. For the most part, the private sector participated in the CDM because it was the first large-scale market mechanism to address climate change and they bought the hype that greenhouse gases represented a new, trillion dollar market opportunity.

That was then. The question for debate and which needs to be resolved is what incentives are required to encourage the private sector to participate at scale in market mechanisms designed to address climate change. The answer to this question is critical because we cannot afford the time that is required to develop a new mechanism that is not successful. Time is running short.

Many of those involved in the first experience with market mechanisms, including some entrepreneurs, lost a lot of money and folded their tents. Although some made money, I assume that most did not. The banking and financial community are more risk-averse today following the economic downturn and they are subject to greater regulatory oversight. Before making a decision to participate in a market that some believe are exotic and devote the resources to do so successfully, past and potential participants would consider the uneven performance of the CDM and carbon markets in general, the bureaucracy that attempted to regulate them, the unwieldy international negotiating process, and the systems' inability to learn from prior experiences and adapt prior to jumping in again. In order to overcome the past, policy-makers must avoid the mistakes that plagued the CDM.

The goal is craft a policy that encourages both entrepreneurial firms like Natsource, that are critical in the formation of new markets and have a high tolerance for risk, and more established name brand firms to participate in new markets. In evaluating new opportunities, firms like Natsource may show up, but because of past experience, putting capital at risk would require significant returns. In order to secure the participation of leading technology and financial firms, policies must provide a transparent and consistent policy framework. These characteristics will enable investors to evaluate market opportunities and risk. We are beginning to see the benefits of this approach in the clean energy space. The private sector will show up if policies are clear and consistent over time.

The CDM Was Overhyped: Its Problems Were Predictable

The CDM was the subject of much hype. Supporters oversold what it could realistically accomplish. Achieving reductions in GHG emissions at lower costs than other options and stimulating innovation would have been sufficient. It should have been viewed only in that context.

The CDM was designed by representatives from over 150 countries. They were predominantly diplomats and environmental policy-makers. They lacked commercial experience and an understanding of financial markets. The US, the country with the greatest experience with environmental markets and the strongest advocate of market mechanisms in the international negotiations, was not significantly engaged in developing the rules to govern the CDM. It was administered by the UN. Given these realities, it should not have surprised participants that the mechanism was not based on commercial principles, that the project cycle would be lengthy and inefficient, that decisions were often arbitrary and opaque, and that it did not provide for acceptable levels of public input.

In addition to the challenges cited above, other time related issues worked against the CDM. The mechanism was initially designed to serve the first commitment period in the KP, which would run from 2008 to 2012. However, in practice it had to ramp up its operation three years earlier than expected in an attempt to serve Phase 1 of the EU ETS from 2005 to 2007. It was under-resourced, short-staffed, and the supporting infrastructure was not yet in place. Many of the KP parties were delinquent in providing their financial contributions to the CDM, increasing its challenges to hire staff and develop the necessary systems to handle a rapidly growing pipeline of projects. There were not an adequate number of DOEs. Given these issues, the criticisms of the CDM's performance in approving too many projects in its early days and the subsequent bottlenecks that emerged in the project cycle, which hindered its performance and created a lack of confidence among its participants, should have been expected.

The CDM Was Hampered by Inconsistent Objectives

This chapter reviewed some of the literature assessing whether the CDM met its two objectives. The analyses concluded that SD and cost effectiveness were competing objectives. The top-down multi-lateral negotiating process is largely responsible for the overly ambitious international climate change policy that was the KP. The negotiating process is often conducted in a circus like atmosphere in which more than 100 countries that are supposed to have equal voices participate and thousands attend. It should be no surprise that the process resulted in overly ambitious policies and programs that included muddied and competing objectives like the CDM, which were included at the behest of countries with entirely different interests and unrealistic expectations.

My view is that the CDM's objectives of SD and cost control were in conflict. Although this is not a wildly popular view and GHG offset projects should do no harm, I believe their overriding purpose should be to reduce emissions at the lowest possible cost. And this is the view of the overwhelming majority of the buyers that are required to spend money to reduce their GHG emissions. This was the primary objective of Natsources' fund participants. Additional programs can be put in place to achieve SD and other objectives.

Because of the need to avoid the pitfalls of the international negotiating process, alternative arrangements between countries and other members of civil society, sometimes called carbon clubs, are becoming more popular models of climate diplomacy. In these arrangements, participants can agree to achieve more limited goals than are the focus of the multi-lateral international negotiations conducted by the UN. These clubs may help to forge more limited, attainable agreements in climate change policy 2.0 than would be possible under the approach that has dominated international climate change policy in 1.0. This is the subject of further discussion in Chap. 6.

Some of the Criticisms Leveled at the CDM Were Unfair

There were many criticisms leveled at the CDM. Several were unfair and hampered its effectiveness for years. One such criticism was that the geographical distribution of projects was unequal with too many centered in China and other large developing countries. China hosted more than 50 % of CDM projects that created approximately 60 % of the CERs by the end of 2012. This criticism was off base. China was the most logical country to host CDM projects. It is the largest developing country economy in the world and provided many GHG emission reduction opportunities. The government created a DNA that provided investors with consistent decision-making reducing one of the many risks that characterized CDM projects. India and Brazil also hosted a large amount of CDM projects for many of the same reasons.

Many argued that the CDM was a failure because of the minimal number of projects that were located in less developed countries (LDCs) to contribute to SD and increase the local populations' access to energy and water. This was not the purpose of the CDM. Although a follow-on mechanism, may be able to make a contribution to these objectives, this was never going to happen at scale given the size of the projects in these countries and investment related risks. At Natsource, our business model was predicated on the need to earn sufficient returns based on the risk of doing CDM business in large developing countries. Investment in less advanced countries would have required even greater returns. Assisting LDCs with their SD and increasing their population's access to energy and other vital commodities and services is as much the responsibility of governments and international institutions as the private sector. But for additional financial incentives, insurance, and other mechanisms that reduce risk, large-scale private sector engagement in LDCs is an impractical objective of a market mechanism designed to address climate change.

Another objection to the CDM that was voiced regularly and which was mentioned previously were criticisms of the prominence of industrial gas projects, particularly HFC-23 projects. Chapter 3 provided my views why these projects benefits outweighed their costs.

The Composition of the CDM Executive Board Contributed to Its Failure

The criterion for being selected as a board member of the EB was based on where you lived. Specifically, out of the ten members, five were to be appointed from the UN's five regional groups, two would be from Annex I countries, two would be from non-Annex I, and one would be from the small island developing states.[132] This makes no practical sense but was

---

[132] *Report Of The Conference Of The Parties On Its Seventh Session, Held at Marrakesh From 29 October to 10 November 2001, The Marrakesch Accords. Addendum, Part Two: Action Taken by the Conference of the Parties, Modalities and procedure for a clean development mechanism as defined in Article 12 of the Kyoto Protocol.* Draft decision -/CMP.1 (Article 12). *Annex C. Executive Board.* United Nations. 2002.

typical of the Kyoto Protocol. There was a lot of concerns about decision-making and inclusiveness and less about substance.

The CDM was an entirely new mechanism. Although a few pilot GHG emissions offset systems had been in place, nothing remotely comparable to the scale of the CDM had ever been attempted. Its board should have been comprised of members with policy, regulatory, technology, and financial expertise and experience running large-scale operations. The composition of the EB damaged the CDMs prospects from the beginning.

The Executive Board Did Not Act Like Executives

The word executive connotes the ability to lead, manage, and administer. The EB's function was supervisory in nature. There was a sense from the overwhelming majority of market participants that the EB did not act in an executive capacity. Rather, it micro managed the mechanism. It second-guessed the DOEs it accredited and ultimately suspended some, leaving project proponents to scramble to find new ones. The project cycle became far more difficult to navigate when this began to occur. They appeared to engage in the nitty-gritty of reviewing specific projects and every document that was created. These actions helped to create the backlog in the project cycle that will be referred to below. And it did not appear that the COP/MOP, which had authority oversee the entire mechanism and to provide guidance to the board, did much to reign in the EBs proclivity to micromanage.

The CDM Project Cycle, Its Administration, and Decision-Making Were Less than Optimal

The CDM project cycle did not work well. Projects entering the pipeline increased from 36 per month in 2005, to 53 in 2006, to 90 in 2007, increasing to an average of 116 projects per month through the first ten months of 2008.[133] The CDM was ill equipped to handle this pipeline due to the significant workload and resource constraints mentioned previously. Bottlenecks appeared consistent with the growth in projects

---

[133] Kossoy, A., P. Ambrosi. *State and Trends of the Carbon Market 2010.* The World Bank. P. 41.

entering the system. The time required for a project to reach registration increased from approximately 370 days in 2007, to approximately 570 days in 2009.[134] In 2007, CDM projects took 316 days to move from registration to first issuance, increasing to over 500 days in 2008, and over 600 days in 2009.[135]

The bottlenecks were ultimately alleviated, but by that time the lack of demand for CERs rendered it a moot issue from a macro perspective. However, delays mattered to investors. This is because many relied on CER revenues for project implementation. Based on our contacts with project developers, many lost interest and could not move forward. Inefficiency and transaction costs are going to hamper offset systems.

In addition to the bottlenecks, market participants believed the EB acted arbitrarily in the project cycle, particularly with regard to the application of additionality. Requests to review a project prior to registration increased from 24% in 2006, to 47% in 2007, to over 60% in both 2008 and 2009.[136] Whereas concerns were voiced over the quality of many projects, this appears to be a very high percentage. Rejected projects increased from 4% in June through December 2006 to 17% in the same time frame just one year later. Over 10% of projects were rejected in 2008 and 2009.[137] Changes were made to methodologies that had heretofore been approved. Baker and Mackenzie, an international law firm, documented 15 major rule changes designed to streamline CDM procedures, address additionality and baselines, and expand the mechanism.[138] The mechanism was always in flux.

Other issues of importance to investors, and which are stables of effective administrative processes, are transparency, consistent decision-making, a process to appeal decisions, and communication. As described in the discussion of the AMPCO project in Chap. 3, the EB made decisions behind closed doors, frequently made arbitrary decisions, and there was no ability for appeals. Decisions were final. In addition, and unlike most

---

[134] Kossoy, A., P. Ambrosi. *State and Trends 2010.* P. 42.

[135] Kossoy, A., P. Ambrosi. *State and Trends 2010.* P. 42.

[136] UNEP DTU Partnership. *CDM/JI Pipeline Analysis.*

[137] Shishlov, I., V. Bellassen. *10 Lessons From 10 Years of the CDM.* CDC Climate Research. N. 37, 2012. P. 10.

[138] Shishlov, I. V. Bellassen. *10 Lessons.* CDC Climate Research. P. 9.

government administered programs, the EB did not communicate directly with market participants. Public officials and those they regulate can benefit from direct communication and interaction because they can learn and better understand each other's concerns and views. However, EB members communicated directly with DOEs which then communicated with project participants. The indirect communication prevented learning and also contributed to delays and bottlenecks because of the need for the DOE to communicate with the EB and then with those making an inquiry.

From a practical perspective, the issues cited above adversely impacted project developers and compliance buyers. The bottleneck in the project cycle and changes in methodologies delayed CER deliveries and impacted firms' ability to meet their CER delivery obligations, requiring them to either default or purchase higher priced CERs in the market to meet their obligations. To repeat a point made previously, private firms make investments based on predictability, certainty, and assessment of risk. The CDM's performance was not predictable or certain and imposed significant risk. This damaged its credibility with stakeholders, reducing its political viability.

Governments Took Many Decisions that Hurt the CDM

The performance of environmental markets and market mechanisms are greatly affected by the governments that design and regulate them. The CDM is no exception. Many actions taken by governments and the EB adversely impacted the CDM's performance. Inaction was also costly. For example, the inability of the international community to make progress on a successor treaty to the KP and agree on the role of markets in the post-2012 period injected uncertainty into the CDM market as to whether there would be a demand for CERs after the conclusion of the 2008–2012 period. This stifled the potential for a post-2012 market when one may have been possible.

The EU took several actions that adversely impacted the market. These included the imposition of quantitative and qualitative restrictions on the use of CERs and ERUs. Quantitative restrictions limited the amount of CERs and ERUs that could be used between 2008 and 2012.[139] The volume

---

[139] According to Ellerman et al. in Pricing Carbon, the EU provided guidance on this issue in 2006 November.

was increased modestly through 2020.[140] And the EU conditioned the use of a higher amount of CERs in the 2013–2020 period on the successful negotiation of a successor treaty to the KP, further increasing uncertainty.[141] This likely chilled project development as investors needed to assess whether additional CER demand would materialize in the world's largest buying region. Qualitative limitations were imposed on credits created by nuclear power, certain types of forestry and later were expanded to industrial gases.[142,143] Although many of these policies had no affect because of the supply and demand imbalance that materialized, these examples illustrate the impacts that government decisions have on of environmental markets.

In this section, I have attempted to illustrate some of the CDMs successes and failures. The CDM likely reduced emissions from what they would have been in the developing world, leveraged large amounts of capital, and lowered firms' and Parties' costs to comply with GHG emission reduction requirements. However, the mechanism had many shortcomings and its successes came at a great cost. The most important of which may have been to reduce governments', investors', and other stakeholders' confidence in global market mechanisms as an effective tool to address climate change. Hopefully, the experience and lessons learned from the CDM and JI will be beneficial in developing the rules to govern market mechanisms under development.

The Kyoto Protocol

The KP, which was negotiated under the UNFCCC framework, was climate change 1.0 at the international level. It was the result of negotiations between nearly 200 countries from 1995 to1997 and rules governing its implementation were negotiated for several years following that.

---

[140] DG CLIMA. *EU ETS FAQ : Question 20*. Available at : http://ec.europa.eu/clima/policies/ets/faq_en.htm [Accessed 9 December 2015].

[141] DG CLIMA. *EU ETS FAQ : Question 20*. Available at : http://ec.europa.eu/clima/policies/ets/faq_en.htm [Accessed 9 December 2015].

[142] European Commission. *Questions & Answers on Emissions Trading: Use Restrictions for certain industrial gas credits as of 2013*. [Press Release]. 25 November 2010.

[143] European Commission. *Emissions trading: Commission welcomes vote to ban certain industrial gas credits*. [Press Release] 21 January 2011.

The KP can be described as top down, overly ambitious, and complex. This is evidenced by its approach to mitigation and administration. Its mitigation model which was negotiated by nearly 200 countries included inflexible GHG emissions limits that failed to recognize trends and differences amongst Annex I parties and the attempt to create a global market to achieve these even though there was little to no experience to date with such programs at any level. The KP's governance and administration was also top down, overseen by nearly 200 countries, and administered by the UN bureaucracy. And in addition to attempting to address climate change, the UNFCCC and the KP was loaded up with other issues of enormous importance including development and equity, adaptation, technology transfer, and trade to name a few which made it more difficult to achieve progress.

International climate change policy 1.0 is over as embodied by the KP and its market mechanisms. Many hold different views as to when that occurred. One could argue it was over before it started, when the 1995 Berlin Mandate[144] was agreed to or it may have been when the Kyoto Protocol[145] was gaveled through two years later at COP-3, or when the Bali Action Plan[146] was adopted at COP-13 which decided to develop a post 2012 agreement. The date does not matter. There has been a clear consensus for many years that a new approach was required to address global climate change.

Although much was accomplished prior to and during the KP period, including the learning by those that participated in its development and implementation, I consider it to be a failure for two over-arching reasons. The first is that, by any measure, GHG emissions grew significantly and GHG concentrations continued to rapidly accumulate in the atmosphere. And perhaps even more problematic was that over 20 years was spent developing and implementing the KP. This time cannot be recouped.

---

[144] *Report Of The Conference Of The Parties On Its First Session, Held At Berlin From 28 March To 7 April 1995, The Berlin Mandate.* Decision 1/CP.1 United Nations. 1995

[145] *Kyoto Protocol to the United Nations Framework Convention on Climate Change.* United Nations. 1998.

[146] *Report of the Conference of the Parties on its thirteenth session, held in Bali from 3 to 15 December 2007, Bali Action Plan.* Decision 1/CP.13. United Nations. 2008.

Much of the discussion on climate change during the 1990s focused on the temporal dimension of the issue. Unlike many important public policy problems that require immediate intervention, many viewed climate change as a century-scale challenge. The objective of the UNFCCC was the 'stabilization of greenhouse gas concentrations in the atmosphere at a level that would prevent dangerous anthropogenic interference with the climate system'.[147] This means that unlike other environmental issues the focus was on the atmospheric concentration of GHG levels which is largely a function of the long-term accumulation of GHG emissions rather than annual emissions. Surely in the time frame that was available, governments and society could put the institutions and policies in place and develop and deploy the necessary technologies to achieve climate policy objectives. Time was available to slow the trajectory of GHG emissions and then reduce them in an orderly fashion in order to minimize the adverse economic impacts that could result from moving to rapidly. The KP with its emphasis on near-term emissions limitations and the efforts to create a global market distracted us from putting the institutions and policies in place to address the long-term nature of the problem.

Unfortunately, time is no longer on our side. Approximately 20 years have passed since the KP was agreed to and much less time exists to take the actions necessary to mitigate climate change and adapt to its impacts. Government delay in responding to climate change is similar to an individual that procrastinates prior to taking action to address an urgent personal or professional matter. There is no time left to waste in fashioning effective and enduring responses to climate change. We are at an urgent point in time.

Much has been written regarding the problems that plagued the KP's structure and performance. There is little utility in repetition. However, it is important to focus on some of the mistakes that led to the KP failures in order to avoid repeating them while creating a more effective follow-on agreement. I remember asking a senior member of the US House Energy and Commerce Committee a question about the prospects of an important piece of legislation. He responded that there were two problems: substance

---

[147] United Nations. *United Nations Framework Convention on Climate Change.* Article 2.United 1992.

and process. I felt optimistic since he said there were only two problems. Then I thought about, and it became obvious that those two areas encompassed everything.

Substantive Issues with the KP

What follows are a brief review of some of the substantive issues that contributed to the KPs demise and which should be avoided in developing a successor agreement.

Lack of Coverage Leads to Increased Emissions and Concentrations

As indicated several times, a fatal flaw with the KP was that it never came close to covering even half of global GHG emissions measured from its 1990 base year. This is because the US did not participate and developing countries did not make reduction commitments. Further, although Canada ratified the KP, the country did not take any actions to reduce its emissions. There are many different data sources that illustrate the KPs shortcomings in this area. I attempt to demonstrate the problem as simply as possible through the data below. I will use GHG emissions data including land use given that this is the way in which the KP was developed.

Global GHG emissions were approximately 34–35 $GtCO_2e$ in 1990 including land use change and forestry.[148,149] Annex I parties accounted for approximately 18 $GtCO_2e$ or more than half of the total amount.[150]

---

[148] CAIT Climate Data Explorer. 2015. Washington, DC: World Resources Institute. Available online at: http://cait.wri.org

[149] The 2014 IPCC Summary for Policymakers indicates that 1990 global GHG emissions were 38 $GtCO_2e$. I chose to use the lower data as all data sources indicate a slight difference in total GHG emissions. If I used the higher amount of emissions for illustrative purposes, the KP would have covered an even lower percentage of global GHG emissions.

[150] This data is from the Annex I Parties data included in GHG profiles under the GHG data section of the UNFCCC website. Available from: http://unfccc.int/ghg_data/ghg_data_unfccc/ghg_profiles/items/4625.php

The US and Canada accounted for between 6 and 6.5 $GtCO_2e$ of the Annex I share.[151] Thus, when the US and Canada's emissions are subtracted from the total, the KP ultimately covered between 11.5 and 12 $GtCO_2e$ or approximately 33 % of total global GHG emissions. By the end of the KP commitment period in 2012, global GHG emissions had increased to approximately 48 $GtCO_2e$, representing an increase of approximately 40 % from the 1990 base year.[152] Annex I parties were responsible for approximately 15 $GtCO_2e$ of this amount.[153] The US and Canada accounted for approximately 6.5 $GtCO_2e$ of the Annex I totals in 2012.[154] Thus, the KP covered 8.5 $GtCO_2e$ or less than 20 % of global GHG emissions by the end of its commitment period.[155] And CO_2 concentrations in the atmosphere increased from approximately 350 parts per million volume (PPMV) in 1990 to nearly 400 PPMV in 2012.[156]

You do not need to be good at math to know that the KP's lack of coverage made it impossible to achieve absolute reductions in GHG emissions or even to slow growth. It was known for some time that the trends were not sustainable. The answer to the problem is obvious. A successor agreement requires global participation. All countries with substantial GHG emissions need to make commitments to reduce them. Global coverage must be the focus of the successor to the KP. Chapter 6 describes the trends in this issue.

---

[151] The range of US and Canada's 1990 GHG emissions including land use and forestry represent the low and high estimates from the CAIT Data Explorer cited above and US and Canada profiles included in the GHG data section of the UNFCCC website.

[152] CAIT Climate Data Explorer. 2015. The data is included in the Historical Emissions Data Section. Washington, DC: World Resources Institute. Available online at: http://cait.wri.org

[153] This data is from the Annex I Parties data included in GHG profiles under the GHG data section of the UNFCCC website.

[154] This is the approximate mid-point of the high and low estimates of US and Canada's GHG emissions including land use and forestry from the World Resources Institutes CAIT Climate Data Explorer and the GHG data section of the UNFCCC website.

[155] This is arrived at by dividing 8.5 $GtCO_2$ (Annex I 2012 total minus the US and Canada) by 48 $GtCO_2$ (the 2012 global total).

[156] Dr. Pieter Tans, *NOAA/ESRL (www.esrl.noaa.gov/gmd/ccgg/trends/)* and Dr. Ralph Keeling, *Scripps Institution of Oceanography (scrippsco2.ucsd.edu/)*

Comparability and Equity

The concept of comparability is an important consideration in the establishment of countries' GHG emissions targets, one that has significant political ramifications. Comparability refers to the extent to which countries' levels of effort in reducing their GHG emissions are similar. This is important because of the perception that firms responsible for achieving a large portion of national GHG emissions targets could be economically disadvantaged if their GHG reduction targets are relatively more rigorous than firms in competitor nations. Simply put, this is because firms in a nation required to undertake greater efforts than competitors in other nations would confront higher costs and potentially lose market share. This could translate to lost economic output, jobs, and income in the country that took on the more rigorous target.

During the negotiation of the KP, some in the US were concerned that the EU was using the negotiations to bolster its economy at the expense of the US. This was because of the EU's position on a series of interrelated issues that included: (i) Article 4 of the KP, which authorized the EC to achieve its KP target of 8% below 1990 levels on an aggregate basis and to distribute the burden to its 15 pre-2004 members as it deemed appropriate; (ii) a belief that reductions in the EU's GHG emissions that were unrelated to climate policy would make a major contribution to the achievement of its target; and (iii) the EU's attempts during the negotiations to limit the use of flexible mechanisms for compliance with GHG emissions targets. At the time, the US was the primary advocate of the use of market mechanisms to achieve climate policy objectives.

The EU BSA or bubble enabled the 15 pre-2004 EU member states to comply with its KP target on an aggregated basis. In other words, the EC would be assigned one numerical target for the purpose of complying with its KP commitments. Each member would then be assigned a target that would roll up in to the overall target.[157] The EU formally adopted the BSA in June 1998 and each member state assumed a GHG emission reduction target.

---

[157] For a review of the EU burden sharing targets see, European Environment Agency, *Questions and answers on… key facts about Kyoto targets*. June 2010.

It appeared hypocritical to many in the US that while the EU was taking a hard line in the negotiations against flexibility and the use of market mechanisms for compliance, the BSA provided the EU with significant flexibility in achieving its target. This is because the EU could meet its overall target even if an individual member state failed to achieve its targets.

The EU allocated targets to member states to achieve its KP target of 92% of base year emissions. Compliance with the target required reductions in the range anywhere from approximately 315–340 $MtCO_2e$ per year.[158] And although several of the 15 countries would be required to achieve reductions, Germany's EU target of 21% below 1990 levels translated to reductions between 230 and 260 $MtCO_2e$ and the UK's target of 12.5% below 1990 levels translated to reductions ranging between 90 and 100 $MtCO_2e$. Combined reductions for German and the UK would be in the range of 320–360 Mt of reductions; which were generally equivalent to the amount required for the EU 15 under the BSA.[159] The way that the BSA worked would result in Germany and the UK achieving the large majority of the reductions required for EU compliance. Ireland, Portugal, Greece, and Spain's targets allowed for GHG emissions growth of more than 90 $MtCO_2e$.[160]

The BSA became more controversial due to a belief that a significant portion of Germany's and the UK's GHG emission reductions would be achieved for reasons unrelated to climate change. Those with this view believed that Germany's GHG emissions baseline was higher because of the reunification of East and West Germany and, correspondingly, emissions would fall rapidly once inefficient power plants and industrial facilities were closed in the east. An analysis concluded that Germany's GHG emissions in 2000 were approximately 225 Mt (18%) lower than 1990

---

[158] The European Environment Agency, *Questions and answers on…key facts about Kyoto targets, 4 June 2010* indicates that the EU 15s Kyoto Protocol target is 341 $MtCO_2$. 2010. The UNFCCC data indicates the target would be similar and the CAIT data indicates a target of approximately 315 $MtCO_2$.

[159] The same data sources as above are used to determine the range of Germany's and the United Kingdom's targets.

[160] European Environment Agency, *Questions and answers on… key facts about Kyoto targets.* June 2010.

levels.[161] Approximately 106 Mt or nearly 50 % of these reductions were attributed to reunification.[162]

The issue unrelated to climate policy that affected the UK's GHG emissions was the increased use of natural gas at the expense of coal resulting from the liberalization of energy markets. The same analysis found that the UKs GHG emissions in 2000 were approximately 90 Mt (12 %) lower than 1990 levels.[163] Approximately, 42 Mt or 45 % of the reductions were attributed the liberalization and privatization of energy markets.[164]

Although both Germany and the UK were aggressive in putting policies in place to achieve their KP targets and have been leaders in the effort to address climate change, events unrelated to climate policy made a major contribution to lowering their GHG emissions and the EU's compliance with its KP target. Germany and the UK achieved reductions in excess of 300 Mt by 2000 from the KP's 1990 base year; which was nearly all of the reductions required by the EU to comply with the BSA. Half of these reductions were unrelated to climate policy. This dynamic was similar to cheap natural gas that resulted from fracking in the US and which contributed to declining emissions in the power sector.

During the period in which the KP negotiations were unfolding, the US was experiencing a period of rapid economic and emissions growth. Analysis undertaken shortly after the KP was finalized concluded that BAU US energy-related carbon emissions would increase by over 30 % from 1990 levels to approximately 1800 million tonnes of carbon or nearly 6.6 $GtCO_2$ in 2010.[165] In order to meet its KP target of 7 % below 1990 levels, the US would have been required to achieve annual

---

[161] Eichhammer, W., U. Boedde, F. Gagelmann, E. Jochem, N. Kling, J. Schleich et al. *Greenhouse gas reductions in Germany and the UK—Coincidence or policy induced.* Federal Environmental Agency. Research Report 201 41 133, 2001. PP. 38–39.

[162] Eichhammer, W., U. Boedde, F. Gagelmann, E. Jochem, N. Kling, J. Schleich et al. *Greenhouse gas reductions in Germany and the UK.* PP. 38–39.

[163] Eichhammer, W., U. Boedde, F. Gagelmann, E. Jochem, N. Kling, J. Schleich et al. *Greenhouse gas reductions in Germany and the UK.* PP. 38–39.

[164] Eichhammer, W., U. Boedde, F. Gagelmann, E. Jochem, N. Kling, J. Schleich et al. *Greenhouse gas reductions in Germany and the UK.* PP. 38–39.

[165] US Energy Information Administration, *Impacts of the Kyoto Protocol on U.S. Energy Markets and Economic Activity.* US Department of Energy. SR/OIAF/98-03, 1998. ES. XV.

reductions of over 540 Mt of carbon or approximately 2 $GtCO_2$ during the KP commitment period from BAU.[166] This was perceived as a lack of comparability.

Negotiations with US Business

The issue of comparability and lack of developing countries commitments played a major role in the US refusal to ratify the KP and its ultimate failure. It was always a long shot that the US would ratify given the politics surrounding the issue including opposition by large segments of the business community and political officials to taking any action to reduce GHG emissions.

It is interesting to look back to this period and ask if it was ever possible for the US to agree on a target that would have been acceptable to some of the key sectors in the business community. There may have been the potential at one time to secure the support of an important sector of the energy industry for a GHG reduction target. Administration officials were engaged in discussions with influential electric utilities prior to President Clinton's speech detailing the US negotiating position prior to Kyoto. These companies were never enthusiastic about a Kyoto-like agreement that included targets and timetables for GHG emission reductions, but the companies were pragmatic: they thought that climate change was going to be addressed, they understood their contribution to the issue, and they were strong supporters of the market-based approach that the US had been advocating. Several companies had developed USIJI projects to gain needed experience with such projects and to create political support for the mechanisms. Some of the CEOs of these companies and several others from leading US companies had participated in meetings with President Clinton on the climate issue in the lead-up to Kyoto. Many of these companies had prospered in the Clinton years and believed that the Administration clearly understood the adverse impacts a bad agreement could have on the US economy and its legacy. Because of this, many had hopes they could reach agreement with the Administration.

---

[166] Energy Information Administration, *Impacts of the Kyoto Protocol.* ES. XV.

Since there was strong agreement on the policy framework that the Administration was advocating, discussions ultimately centered on the stringency of the GHG emissions target. To date, the Clinton Administration had been publicly silent on the issue. However, it was understood that 1990 would be used as the baseline to establish the target. In the discussions, some in the industry advocated for two five-year budget periods. The first would require 1990 emissions levels from 2010 to 2015 and some level of additional reductions would be required from 2015 to 2020.

Agreement was never reached on this issue. Those in the industry interested in reaching an agreement could not agree amongst themselves on the target because their fleets comprised power plants of different vintages. Because of this, the targets would have had different impacts on the companies. Some of the utilities also did not want to agree to a target, which they had always opposed in principle. In addition, the Administration could not be bound to an agreement in the subsequent negotiations. In his speech detailing, the US position before the final Kyoto negotiations, President Clinton committed to returning US emissions to 1990 levels from 2008 to 2012 and to further reductions after that.[167]

The US ultimately agreed to a GHG reduction target of 7% below 1990 levels in Kyoto. This was far more stringent than the target proposed by President Clinton in his speech and discussed with industry prior to it. For the most part, the target combined with some of the other issues discussed in this section strengthened industry and political opposition in the US to the KP. Some felt betrayed by the Administration, believing that it had agreed to much more stringent reductions than were necessary to get an agreement.

Hot Air

Hot air is another well-known component of the KP that has been written about extensively. Generous targets were provided to several countries in the KP, particularly Russia and Ukraine. Their actual emissions and

---

[167] William J. Clinton: 'Remarks at the National Geographic Society,' October 22, 1997. Online by Gerhard Peters and John T. Woolley, *The American Presidency Project*. http://www.presidency.ucsb.edu/ws/?pid=53442.

BAU estimates were well below their 1990 baseline due to the major contraction in economic activity following the collapse of the Soviet Union. Russia and Ukraine's GHG emissions were more than 50% lower in 2000 than in 1990.[168] This meant that Russia and Ukraine would have large supplies of surplus units to sell in the international market that would be established by Article 17.

At the time, I thought this was a good idea. In particular, the market would be short and the US, Canada, and Japan would have a large demand for GHG emission reductions. The surplus Russian and Ukrainian supplies could provide liquidity into the market. This would be necessary at the beginning to build confidence, control compliance costs, and provide a source of revenue for sellers to modernize their infrastructure. However, many believed that these countries should not be provided a monetary reward because they did not take any actions to reduce their GHG emissions. Without providing my views whether this made sense from a policy perspective, the result of the subsequent debate was to delegitimize international emissions trading and increase political opposition to it. Although initiatives such as green investment schemes were put in place in an attempt to ensure that revenues from sales of this surplus were devoted for specified purposes, the international trading market never matured.

An Over-Reliance on Markets Was a Problem

Each sovereign country that was a Party to the KP had the ability to establish its own policies to achieve the reductions required to comply with its GHG emissions target. In practice, the global community spent an enormous amount of time and political capital to include the market mechanisms in the KP with the goal of creating a global carbon market. And because so much was invested in this, the incentives were for countries to develop systems which would link to the global market to secure the economic benefits that were being advertised. After all, a significant body of research concluded that the provision of flexibility in the timing

---

[168] Russia's and Ukraine's GHG emissions are from the GHG emission profiles for Annex I Parties and major groups on the GHG data section of the UNFCCC website.

and location of reducing GHG emissions would achieve significant cost savings compared to alternative policies.

Betting on a global carbon market was a risk. However, the US had indicated that its participation in a global regime was contingent on a robust market-based approach. And any agreement that did not include the world's largest economy and emitter at the time was doomed to fail. So, this is the direction that was taken. The risk in utilizing this approach was created by the lack of experience with such mechanisms in general and GHGs specifically. The CAAA of 1990 created a market for SO_2 emissions in order to mitigate acid rain. It was the largest environmental market to date and was cited by its supporters as evidence of the model's success. Yet, although the program was achieving significant results, Phase 1, which ran from 1995 to 2000, had only been in existence for two years when Kyoto was adopted. More importantly, the characteristics of the market for SO_2 emissions were entirely different from a carbon market in several important respects. Some of the differences are highlighted below.

Geographic Scope

The acid rain program was national in scope. In contrast, Articles 6 and Article 17 of the KP creating JI and the international emissions trading program would necessarily involve all Annex I countries. At least a hundred countries would participate in the CDM.

Coverage of Sources and Gases

The acid rain trading program covered the sulfur dioxide emissions of approximately 110 power plants for its first five years in states primarily east of the Mississippi River. The program would be expanded to cover additional plants in Phase 2 beginning in 2000, ten years after the CAAA was adopted. In contrast to the coverage of a minimal number of sources covered in the US program, a carbon market would need to regulate millions of emissions sources and carbon sinks in many countries because

GHG emissions are inherent in nearly every aspect of economic activity in all countries. In addition, the acid rain program covered one pollutant while a GHG program could theoretically cover all of the GHGs covered by Kyoto.

Program Infrastructure

The environmental integrity of a trading program is contingent upon a reliable system in which emissions are monitored, reported, and verified and transactions are tracked in a registry and transparent. The EPA, which already had significant experience regulating the power plants covered by the sulfur dioxide program, was given clear authority in the CAAA to oversee these functions. And regulated sources were required to utilize specific technologies to monitor their emissions and to report them. Each transaction of SO_2 emissions was reported to a registry.

There were stark differences between the US SO_2 program and the carbon market. Each Annex I country would be required to develop the rules and systems necessary to regulate national carbon markets and their firms participation in the global market. They each had different capabilities to do so and likely less than existed in the US. The rules governing the international market would be the result of multilateral negotiations involving many countries and would be supervised by the UN, which was far less experienced than the US EPA. And many developing countries with no experience with environmental markets and minimal regulatory infrastructure would participate in the CDM.

There were few similarities between the US acid rain program and the global carbon market, regardless of advocate's comparisons, including myself at the time. The models could not have been more different.

Because of all of these issues, it should have been expected that establishing a carbon market would be far more complex than what was experienced in the US acid rain program. Reliance on one approach to achieve climate policy objectives was an enormous risk. The challenges should have been anticipated. The surprise is that so much was accomplished in such a short amount of time.

The other problem with the over-reliance on the market to achieve GHG emissions reduction targets was the lack of emphasis in developing other solutions. A former colleague of mine used to say that there was no silver bullet to solving the climate problem. Large-scale reductions in GHG emissions would only result from a portfolio of policies. Because CO_2 emissions are predominantly caused by the combustion of fossil fuels to create energy, one policy that should have received far greater focus during this period was increased development and deployment of low and non-emitting energy technologies. Technological progress is the result of many factors but research development and demonstration (RD&D) is a critical one. Yet, these efforts were deemphasized in the period in which the KP was being negotiated.

Another lesson from the KP, which is discussed in the recommendations section, is the need for diversity of policies in climate change 2.0 that are designed to attack the multitude of the causes of climate change and can be dialed up or down based on their successes and failure.

Temporal Dimension of the Issue

As mentioned previously, climate change is a century-scale issue. Yet, the policy embedded in the KP did not reflect this consideration. The argument was always about near-term targets. And although I understand the need for elected officials to be able to point to accomplishments that were achieved on their watch; the emphasis on GHG emissions targets in 2008–2012 did not make sense, particularly for the US, because of the economic expansion that was occurring. The US target required too many reductions in a short period of time.

It would have been sufficient to agree on near-term targets that would begin to slow emissions growth and establish goals for the mid and long term. This would have provided an incentive to avoid investment in long-lived carbon-intensive technologies while providing the regulated community and investors with a signal that climate policy was here to stay.

The Lack of Long-Term Certainty

Significant reductions of GHG emissions are required throughout the twenty-first century to mitigate the risks of climate change. Achieving them requires dramatic improvements in carbon and energy intensity. The private sector will be vital to this effort. It is a source of capital and has the ability to bring lower emitting GHG technologies and processes to the market. Government can secure the sustained commitment of the private sector to address climate change by putting a consistent, enduring policy framework in to place. This provides certainty, and the market signals that are necessary for business to evaluate and make the long-term investments that will be required throughout the twenty-first century.

The policy framework embodied by the KPs five-year, near-term emissions limit did not convince a large contingent of the private sector that governments were serious about climate change. There was great uncertainty as to what, if anything would come after 2012. The commitment was not there. This shortcoming of the KP needs to be avoided in the attempt to create a durable successor agreement.

The International Negotiations Process Served Too Many Masters

This is a similar issue to one that plagued the CDM's performance. It also adversely affected the UNFCCC and KP. The climate negotiating agenda got overloaded with issues that are of great consequence in their own right. I have a hard time imagining how multi-lateral negotiations involving nearly all of the world's countries can successfully address such issues as emissions mitigation, adaptation, finance, technology transfer, economic development, sustainable development, equity, and poverty alleviation to name but a few. All of these issues in their own right are enormously important and require new solutions. The process used to negotiate the UNFCCC and KP was not capable of addressing all of these issues.

Process-Related Issues

This section focused on some of the substantive issues that contributed to the KPs downfall. The substance embedded in the KP was an outcome of a flawed process. Looking back almost 20 years, it seems incredible that the Kyoto Protocol was negotiated in a two years. Given what was at stake, and the complexity of the agreement, this alone seems to be impossible. What follows are a few issues related to the process that needs to be avoided in developing the rules necessary to implement the Paris agreement.

Universal Participation and Decision-Making

This issue is the focus of greater discussion in Chap. 6 regarding the negotiation of the follow-on agreement and activities that are influencing it in climate change 2.0. The KP process included nearly 200 countries. I understand the need for inclusion and for all countries voices to be heard. However, in 1990, the US, EU 15, China, Russia, and Japan accounted for approximately 65 % of the world's 1990 CO_2 emissions from fuel combustion and approximately 50 % of global GHG emissions.[169,170] Their views on issues of importance should have been provided greater weight than less consequential players. This appears to be the view of larger emitters that are increasingly making agreements outside of the ongoing international process. The substance of these agreements, and their potential impacts on international climate change 2.0, are discussed in greater detail in Chaps. 5 and 6.

A subset of participation is decision-making, which is supposed to be unanimous according to UN rules. This is not possible given the disparity of views on the climate change issue. At times, the Chair has gaveled through decisions even though objections were raised. However, at other

---

[169] IEA (2014), CO_2 Emissions From Fuel Combustion 2014, www.iea.org/statistics © OECD/ IEA, Paris, IEA Publishing. Licence: www.iea.org/t&c

[170] CAIT Climate Data Explorer. 2015. Washington, DC: World Resources Institute. Available online at: http://cait.wri.org

times countries have been able to slow and even halt progress because of the decision-making requirement.

The Process Was a Zoo

I was amazed that anything could be achieved in COP meetings. The only analogy to what I witnessed in the COPs, and it is not a perfect one, is a mark-up of tax or other comprehensive legislation in the US, such as the CAA. Since I have participated in those, we can use that as the example. During a mark-up, members of the relevant committees go through the bill and consider amendments to it and vote on them. This occurs before the full House of Representatives or the US Senate votes on legislation. Hundreds of lobbyists' mill around outside the committee room as the mark-up occurs. It is not uncommon for staff or a committee member to ask a lobbyist representing an interest from their congressional district or state whether an amendment would be acceptable and what the local impacts could be. The presence of such large groups of affected parties complicates the decision-making process.

The international climate change negotiating process is a mark-up on steroids. Thousands of representatives from business, NGOs, and others mill about for as long as two weeks at COP meetings as delegations meet in large and small meetings in an attempt to resolve complex and contentious issues. It is extremely difficult to conduct business and make decisions in this fashion.

The Participation of Heads of State Prolongs Inadequate Processes

At COP meetings, Heads of State frequently fly in to break logjams in the negotiations and to celebrate successful outcomes. Their presence makes it difficult to make a realistic assessment of the negotiation's progress and whether alternative approaches would be beneficial. This is because leaders like success, and do not like admitting to failure. So significant effort is made to cobble together face saving agreements, with great fanfare, even when little or no progress has been made. Leaders celebrate the outcome. Then it is left to staff to attempt to devise a path forward. This consumes an enormous amount of time and resources.

I believe this phenomenon has perpetuated the failures of the UNFCCC process. It is hard to walk away from agreements in the existing process that provide the appearance of success, regardless of how meager they are. We would be a lot further along in 2015 if some negotiators had the audacity to admit that the process was not working and alternative approaches were required.

Heniminvel ius nobis ex eveliqui intionetur min nam nim vel ium im dolumqui qui sum que omnis autatibusa corrum id ut aut aditatur as et, sum dolorit officie ndisciendi dist, iliqui dite similluptat et autatem porrum ime offictem quos assita consequid moluptatius maximi, odias sumque omnissuntio modit debis inctempos moluptasimi, si derovit excea voleceatur am qui amenemp oremporem illorit mo dolorepera doluptatur, cusdam dolupitiur as dunt lit acia que dolentia sum il is reptatem ipsa dis a que atum qui ut dolore volum as exped eum et mil ium haribus nos iligend aeribus, ut es aut qui di aut aceatur ma nihitam vid et ute pelectur sitatia ntemolendis antin et eumendu ntione imi, tem. Facim qui ducipsamet aut delliqui optat pos aute is sitas moloritae. Officae dollabo riostinverat quo moloratios a pariti te etur ab is endiati tem aut et a ne periam, se suntur si dit acerund emperumet quaecum iunt, sitem. Et modit facerspellut essim incil idus mod ut aut officiet re nonem commodi dolo tem. Nem aut utatur mod qui odignam fugiti solest quat fugit, susam vernamus, con corporition praecte cor aut quuntis reperferum comnis suntium repe nectem hillest atemperibus dolessi comniandes dolupta dia prae doluptasim quae necto te molorem eaquas molest expelleces restios alis moluptate cusaero ea qui ius, nistiosant et, ipicabori de volo molore odipien dandipsam faccusa nducipsa sime dolo milit volo estiae aut ommodi berovit dolupta nonsedis et vendi dolupti omnias dolum am, aborpos dis resequodis nam volupta tincim illuptatis minceniat.

La eaturem que pelissit am voloris excepudi core volore nusciis duciuri bustincti cullessum rem quunt rem quidia que magnatem harum et officatent fugia necatur? Cipsuntem voloribus, consedit, quamus disciis imincit ibusam a nate venest, nus re sita nes dus dendae quiam harit, ni

5

The End of Climate Change 1.0 in the US

A Confluence of Events Provide the Impetus for US Action

This chapter reviews the last major effort in the US in climate change policy 1.0; the failed attempt to adopt an economy-wide GHG cap-and-trade system in the period 2009–2010.

Several events prior to and in 2008 created the impetus for the US to develop a domestic climate change program. These include the election of Barack Obama as president, increased public awareness of climate change, the Supreme Court's decision in Massachusetts versus EPA, state-based climate policies, and political dynamics.

President Obama Takes Office

Barack Obama was inaugurated as the 44th President of the US on 20 January 2009. His historic inauguration as the first African American President of US was, in itself, an inspiring source of optimism for many

© The Author(s) 2016   **153**
R.H. Rosenzweig, *Global Climate Change Policy and
Carbon Markets*, Energy, Climate and the Environment,
DOI 10.1057/978-1-137-56051-3_5

in the USA and around the world. When watching his acceptance speech, I remember thinking that he was a different kind of politician than I had ever seen. He appeared to be a transformational leader, something the nation required following the public's increasing weariness with US international engagement after the events of 11 September 2001 and the most serious recession since the Great Depression. As a Senator, Obama opposed the Iraq War, which had become extremely divisive and unpopular, and he promised to end it once in office. He also ran on a progressive domestic agenda promising an ambitious stimulus program to revitalize the economy. President Obama committed to overhaul the US health care system to make insurance coverage affordable for millions of Americans without it, a goal that had eluded every Democratic president for half a century, and vowed to reform the regulations governing the nation's financial system.

With respect to energy and climate, his plans were ambitious. He called for creating five million green jobs, investing $150 billion in clean sources of energy over a ten-year period, increasing the share of renewables to 25% of national electricity supply, and to implement an economy-wide cap-and trade system that would reduce GHG emissions by 80%. He also proposed to auction 100% of the permits.[1]

The potential to advance this agenda would be greatly impacted by deteriorating economic conditions. Approximately two million jobs were lost in the last four months of 2008.[2] And approximately 4.7 million more jobs were lost in 2009.[3] This dictated that the president's first course of business was to develop a program to create the conditions for economic recovery and put people back to work. It also raised the question as to whether the public had the appetite for a far-reaching climate change control program given its potential costs and regional impacts. At the end of January 2009, and without one Republican vote,

---

[1] Obama08. *Blueprint for Change.* Available from: https://ia801003.us.archive.org/19/items/346512-obamablueprintforchange/346512-obamablueprintforchange.pdf [Accessed 17 December 2015].

[2] Bureau of Labor Statistics. *The Employment Situation: December 2008.* [Press Release]. 9 January 2009.

[3] Barker, M., and A. Hadi (2010) 'Payroll employment in 2009: Jobs losses continue', *Monthly Labor Review*, 23–33.

the house passed an \$800 billion economic stimulus package known as the American Recovery and Reinvestment Act (ARRA). It won senate approval nearly two weeks later with the support of three Republicans. It was signed into law on 17 February 2009. The legislation made a down payment on the president's clean energy agenda including \$80 billion for clean energy.[4] However, the narrow margin of victory in passing the legislation foreshadowed the inability of the political parties to work together over the next several years.

Climate Change Awareness Grows in US

Prior to 2008, several factors contributed to an increased awareness of climate change in the US and growing support for addressing the issue.

Intergovernmental Panel on Climate Change Fourth Assessment Report: Climate Change 2007 (AR 4)

The Intergovernmental Panel on Climate Change (IPCC), widely viewed as the world's most authoritative scientific body on climate science, provided scientific support to the UNFCCCs efforts on climate change. It issued its fourth Assessment Report (AR 4) in 2007.[5] AR 4 concluded that 'most of the observed increase in global average temperatures since the mid-20th century is *very likely* due to the observed increase in anthropogenic GHG concentrations'.[6] And it pointed to the role that human activities had in increasing GHG emissions and atmospheric GHG

---

[4]Vice President Biden. *Progress report: The Transformation to a Clean Energy Economy.* Available from: https://www.whitehouse.gov/administration/vice-president-biden/reports/progress-report-transformation-clean-energy-economy [Accessed 17 December 2015].

[5]IPCC, 2007: Climate Change 2007: The Physical Science Basis. Contribution of Working Group I to the Fourth Assessment Report of the Intergovernmental Panel on Climate Change [Solomon, S., D. Qin, M. Manning, Z. Chen, M. Marquis, K.B. Averyt, M. Tignor and H.L. Miller (eds.)]. Cambridge University Press, Cambridge, United Kingdom and New York, NY, USA, 996 PP.

[6]IPCC, 2007: Climate Change 2007: Synthesis Report. Contribution of Working Groups I, II, and III to the Fourth Assessment Report of the Intergovernmental Panel on Climate Change [Core Writing Team, Pachauri, R.K and Reisinger, A. (eds.)]. IPCC, Geneva, Switzerland, P. 5.

concentrations.[7] The report's findings, particularly those regarding human activities influence on the climate system and the potential impacts of climate change, were widely communicated in the mainstream press.

Former Vice President Gore Educates the Public About Climate Change

Former Vice President Gore was engaged full time in the effort to educate the public on the climate crisis, as he called it, and the related risks. He developed a slide show that was turned into *An Inconvenient Truth*, directed by David Guggenheim, which won an Oscar for best documentary in February 2007. The film grossed a combined $45 million in the US and globally, making it, according to Reuters, the third largest grossing documentary ever, except for some concerts and IMAX films. It also sold 24 million DVDs.[8]

In the press release announcing the winners of its prize in 2007, the Nobel Prize Committee said of former Vice President Gore, '[h]e is probably the single individual who has done the most to create greater worldwide understanding of the measures that need to be adopted'.[9]

Former Vice President Gore shared the Nobel Peace Prize with the IPCC in 2007. According to the Gallup environment poll, the percentage of the US public during this time that said the 'greenhouse effect' or 'global warming' 'worried them a great deal' increased from 26% in 2004 to 41% in 2007.[10] It is not clear the increased awareness was caused by

---

[7] IPCC, 2007: Climate Change 2007: Synthesis Report. Contribution of Working Groups I, II, and III to the Fourth Assessment Report of the Intergovernmental Panel on Climate Change [Core Writing Team, Pachauri, R.K and Reisinger, A. (eds.)]. IPCC, Geneva, Switzerland, P. 5.

[8] S. Gorman. *Gore's Inconvenient Truth wind Documentary Oscar*. Available from: http://www.reuters.com/article/2007/02/26/us-oscars-gore-idUSN2522150720070226. [Accessed 17 December 2015].

[9] The Nobel Peace Prize for 2007 to the IPCC and Albert Arnold (Al) Gore Jr.—Press Release. *Nobelprize.org*. Nobel Media AB 2014. Web. 17 December 2015. Available from: http://www.nobelprize.org/nobel_prizes/peace/laureates/2007/press.html [Accessed 17 December 2015].

[10] Skocpol, T. Naming the Problem: What It Will Take to Counter Extremism and Engage Americans in the Fight Against Global Warming: Prepared for the Symposium on: *The Politics of America's Fight Against Global Warming, Co-sponsored by the Columbia School of Journalism and the Scholars Support Network, 14 February 2013*, Report Commissioned by the Rockefeller Family Fund in conjunction with Nick Lehman, Columbia School of Journalism. P. 72.

An Inconvenient Truth and AR 4; however, they likely contributed to this increase. Public opinion is pivotal to passing legislation on an issue as complex as climate change.

Simmering Policy Issues Bring Climate to the Forefront

At the same time public concern increased, the Supreme Court decision and state policy increased the potential for federal action.

Massachusetts Versus Environmental Protection Agency

Chapter 3 described the 2007 Supreme Court decision that ultimately provided the impetus for future EPA regulation of GHG emissions. With Obama coming into office, many believed that the Administration would use this tool to address climate change. This chain of events fueled increased business support for federal legislation to control GHG emissions.

Business generally prefers to address complex public policy issues through legislation than regulation for process and substantive reasons. In the legislative process, business and other stakeholders have significant access to members of Congress. This provides them with the ability to provide input into the process, increasing the potential for a favorable outcome. In contrast, the administrative process that creates regulation is subject to control by experts in the executive branch and governed by strict rules restricting communication. Many in the business community see the process as a black box, particularly when an administration viewed as environmentally sensitive is in office. In the administrative process, the usual remedy to change or eliminate rules is to take the issues to courts after the fact. Litigation is frequently a lengthy and resource-intensive process with highly uncertain outcomes.

From a substantive perspective, legislative solutions also appear to provide greater clarity and flexibility over the long term than rulemakings. This is important for business planning. Rules are frequently viewed as rigid, complex, and often characterized as command-and-control. And, as communicated in an article that focused on the attempt to pass climate change legislation in 2009–2001, rulemakings 'lack the symbolic appeal and democratic legitimacy of legislation'.[11] Congressional actions give legitimacy to an issue and provide stakeholders with a vested interest in fashioning workable solutions.

State Legislation Increases Business Support of Federal Action

As indicated in Chap. 3, in the absence of federal policy, several states intervened on climate change on their own. This included the passage of the Global Warming Solutions Act by California[12] and the establishment of RGGI.[13] In addition, five Western states signaled their intent to develop the Western Climate Initiative (WCI), a program similar to RGGI. It expanded to seven states and four Canadian provinces through 2008.[14] All states, with the exception of California, have since dropped out of the program.

It is not uncommon for states to address environmental concerns by putting policies in place prior to the federal government taking action. An example is acid rain. Five states passed laws or imposed regulations requiring reductions of SO_2 and NOx prior to federal legislation.[15] Due to concerns with potentially higher costs resulting from the need to comply with overlapping rules and conflicting laws, businesses frequently support a uniform federal approach when this situation occurs.

Many businesses concerned with California, RGGI, and WCI programs supported a national solution to climate change. If WCI was adopted, 12 states, including the largest economy in the US would have

---

[11] Layzer, J. (2011) 'Cold Front: How the Recession Stalled Obama's Clean Energy Agenda' in Skocpol, T., L. Jacobs (eds.) *Reaching for a New Deal: Ambitious Governance, Economic Meltdown, and Polarized Politics in Obama's First Two Years* (New York, NY. Russell Sage Foundation). PP. 321–385.

[12] Nunez, F., F. Pavley. *California Global Warming Solutions Act of 2006.* Available from: http://www.leginfo.ca.gov/pub/05-06/bill/asm/ab_0001-0050/ab_32_bill_20060831_enrolled.html [Accessed 4 December 2015].

[13] Regional Greenhouse Gas Initiative. *Overview of RGGI CO$_2$ Budget Trading Program.* Available from: http://www.rggi.org/docs/program_summary_10_07.pdf [Accessed 4 December 2015].

[14] Western Climate Initiative. *History.* Available from: http://www.westernclimateinitiative.org/history [Accessed 17 December 2015].

[15] Garland, C. (1988), 'Acid Rain over the United States and Canada: The D.C. Circuit Fails to Provide Shelter Under Section 115 of the Clean Air Act While State Action Provides a Temporary Umbrella', *B.C. Envtl. Aff. L. Rev. 1*, Volume 16, issue 1, 1–37.

been covered by a GHG control program. There was the strong potential that the programs would be uncoordinated, with different rules, resulting in fragmented, inefficient policies. Companies with assets in multiple states could be regulated by multiple programs and be subjected to different rules and emission reduction requirements. This would increase their compliance costs compared to a uniform federal program.

By 2008, the increased potential for regulation of GHG emissions because of the Supreme Court decision, combined with nascent state policy-making efforts, and ongoing piecemeal implementation of the Clean Air Act, led many in the business community to seek federal climate change legislation. This provided the best opportunity for the certainty many were seeking.

Political Dynamics Increase the Potential for Federal Action

Political dynamics in the US also conspired to create support for federal legislation. These included the creation of a historic coalition of business and environmental interests and Democrats increasing their majorities in the house and senate in the 2008 elections.

The US Climate Action Partnership

Billed as a business and non-governmental organization (NGO) partnership, the US Climate Action Partnership (USCAP) was established in 2007. It advocated for the passage of federal legislation that would 'slow, stop and reverse the growth of GHG emissions over the shortest period of time reasonably achievable'.[16] At its height, USCAP was comprised of approximately 30 members.[17] These included some of the most prominent US environmental organizations such as Environmental Defense

---

[16] United States Climate Action Partnership. *A Call for Action, Consensus Principles and Recommendations*. 2007, P. 7.

[17] The members are listed on the last page of USCAPs, *Blueprint for Legislative Action* which was released in January of 2009 to coincide with President Obama's inauguration.

Fund, a pioneer in developing market-based solutions, Natural Resources Defense Council, The Nature Conservancy, Pew Center on Global Climate Change (now C2ES), and World Resources Institute. Members also included some of the largest and well-known US corporations such as Alcoa, ConocoPhillips, Dow Chemical, Ford, and General Motors. It also included Duke Energy, one of the world's largest coal-fired power generators and CO_2 emitters.

It is rare in Washington to see coalitions comprised of such diverse, prominent stakeholders who are often at odds on public policy issues. When traditional adversaries like environmentalists and business come together to find common solutions to issues, as they did in USCAP, they provide cover for policy-makers to take on contentious issues like climate change. If such a diverse group can agree on the details of policy, it can make the job of finding solutions and achieving consensus easier for policy-makers. There is less friction in the system.

My job at DOE was easier when outside interests were able to reach agreement on an issue or attempt to do so. Although government officials will not usually rubberstamp an agreement reached by affected interests, it provides policy-makers an understanding as to the policy parameters that may be acceptable to those with a stake in an issue. Given the composition of USCAP, members of Congress and the new Administration had to believe that the potential to pass climate change legislation had increased.

The 2008 Congressional Elections

Democrats had a good year in 2008. They added to their majority by picking up 21 seats in the House of Representatives and controlled the house with a 257–178 margin. The Democrats also picked up eight senate seats providing a 57–41 margin. In addition, two independents, Senators Joe Lieberman (I-CT) and Bernie Sanders (I-VT), caucused with the Democrats, effectively expanding the majority to 59–41, depending on the issue. Because Democrats were more committed to addressing climate change than their Republican counterparts, the election's results also increased the momentum to address the issue.

For all of the reasons cited, it appeared to many that the years 2009–2010 provided the best opportunity to date for the US to develop a national program to address climate change. The leadership of a President dedicated to addressing the issue, combined with large congressional majorities, a population with growing concern with climate change, increased business support, and a historic coalition all combined to create optimism that the 15-year deadlock in US policy could finally be broken. The pieces were in place, or were they?

Inside Baseball in the House Impacts the Prospects for Legislation

The scene was set to enact climate change legislation. The process would necessarily entail moving legislation through the House Energy and Commerce Committee before a final vote in the house. In the immediate aftermath of the 2008 election, the question was who would lead that process? At the time of the election, Congressman John Dingell (D-MI) chaired the committee. With the exception of a few years when the Democrats were not the majority party, Dingell had been the committee Chair since 1981. He was one of the old bulls, a term used to describe the powerful committee chairmen that had great influence in Congress. Dingell served in the Congress for 59 years, from 1955 to 2015, longer than any other American in history. He was known as a bread-and-butter Democrat: liberal on economic issues, he was also a supporter of gun rights and could be conservative on national security. Mr. Dingell was known as one of the great legislators in the history of the House of Representatives and he had a hand in virtually every prominent piece of energy and environmental legislation that became law during his tenure.

Under his leadership, the Energy and Commerce Committee was viewed as one of the most powerful in the Congress. Dingell had once quipped that his committee's jurisdiction could be viewed in the art hanging on the wall of the committees' offices: it was a picture of the planet. Importantly, he was able to work with Republicans and forge coalitions to move complex legislation through the committee. He was known as a stickler for detail and maintaining the integrity of the legislative process. As someone that was involved as a lobbyist in the debates over the CAAA of 1990 and the Energy Policy Act (EPACT) of 1992, and worked with the committee frequently while serving as DOE Chief of Staff from 1993 to 1996, I was often in awe of the way he controlled the committee processes and moved complex legislation.

However, Dingell represented the big three automakers in Michigan and was known for his opposition to increased automobile fuel economy standards. As a result, many environmentalists disliked him, despite the fact that he had come around on the need to address climate change. For example, he circulated draft legislation with Congressman Rick Boucher (D-VA) that would have created a GHG cap-and-trade program in the prior Congress.[18] The environmental community viewed Dingell with suspicion.

Right after the 2008 elections, Congressman Henry Waxman (D-CA), who had been elected in 1974 as a member of the reform minded Watergate class and was a long-serving member of the Energy and Commerce Committee, exercised his prerogative to challenge Dingell for the committee chairmanship. He prevailed in a vote of the Democratic caucus on 20 November 2008. Many of the recently elected Democrats supported Waxman, as did many more concerned with environmental issues. Waxman, who was viewed as an effective liberal legislator that championed environmental causes, frequently butted heads with Dingell over increasing fuel economy standards. Important for the climate change issue, he was also viewed as more partisan than Dingell on energy and environmental legislation and did not appear to possess the same relationships with Republicans on the committee or in the full house. The question was now answered: Waxman would be leading the effort to craft comprehensive climate change legislation that could pass the house and ultimately be signed into law.

---

[18] See Pew Center on Global Climate Change for a summary of the 2008 Dingell Boucher discussion draft. Available from: http://www.c2es.org/docUploads/Dingell-BoucherSummary.pdf [Accessed 17 December 2015].

Chairman Waxman would be joined in the effort by Congressman Ed Markey (D-MA), another leading liberal member of the house, who was serving as Chairman of the subcommittee on Energy and Environment and the Select Committee on Energy Independence and Global Warming. Markey, also was a member of the Watergate class, a long-standing member of the Energy and Commerce Committee and viewed by many as one of the most liberal members of the house.

The committee would need to move fast. President Obama's goal was to exhibit leadership at the COP-15 meeting in Copenhagen, which would take place at the end of the year, and had been tasked two years earlier by the Bali Action Plan[19] to establish a successor agreement to the KP. Leadership could only be shown by making progress on comprehensive climate and energy legislation at home.

The Arc of Climate Change Legislation

For those not familiar with the US legislative process, it has been compared to making sausage. The analogy being that if you saw sausage being made, you would not ever want to eat it. The same is true of legislation. You may value the outcome, but the process of creating and passing a complex bill is anything but noble. To those unfamiliar with the ways of Washington, the legislative process can appear to be dysfunctional. In truth, it is often downright ugly, contentious, and uneven. There are always twists and turns. Hundreds of lobbyists with competing views and a lot of money behind them are engaged in the process of pleading their case to members of Congress and their staff. Gaining the support of wavering members of Congress needed to secure a majority for passage of legislation often requires the inclusion of special favors to secure their vote. All of this is magnified in the attempt to tackle an issue as complex as climate change.

---

[19] *Report of the Conference of the Parties on its thirteenth session, held in Bali from 3 to 15 December 2007, Bali Action Plan*, Decision 1/CP.13. United Nations. 2008

US Climate Action Partnership Blueprint for Legislative Action

USCAP released their Blueprint for Legislative Action (Blueprint) on 15 January 2009.[20] Chairman Waxman took the opportunity to hold a hearing on it, at which he set a goal of passing legislation that would address global warming and make the US energy independent by the Memorial Day recess, which was approximately four months away.[21] Fourteen members of the USCAP coalition testified at the hearing, including nine leaders from business and five from the NGOs.[22] The hearing was acrimonious and partisan.

Some USCAP members were concerned that a hearing focused solely on their Blueprint and convened by Chairman Waxman would have the unintended effect of aligning the coalition with the Democrats, increasing the partisanship of the issue. This was not the way to pass complex and contentious legislation. Nikki Roy of the Pew Center on Climate Change said, 'When we didn't insist on pushing off the hearing, we lost a lot of our credibility as a bipartisan initiative.'[23] One of the most important advantages of such a wide-ranging coalition should have been the ability of its business members to work effectively with Republican members of Congress. This appears to have been squandered before the process even began.

Two things struck me about the USCAP Blueprint. The first was its level of detail. Although it was a relatively short document, it called

---

[20] United States Climate Action Partnership. *A Blueprint for Legislative Action.* 2009.

[21] Waxman, Representative H., A. *Opening Statement on the USCAP Legislative Blueprint.* Committee on Energy and Commerce. 15 January 2009.

[22] For a list of witnesses, see Full Committee Hearing on the US Climate Action Partnership, (15 January 2009). Available at: https://wayback.archive-it.org/4949/20141223181331/http://democrats.energycommerce.house.gov/index.php?q=hearing/full-committee-hearing-on-the-us-climate-action-partnership-January-15-2009 [Accessed 17 December 2015].

[23] Bartosiewicz, P., M. Miley. The Too Polite Revolution: *Why the Recent Campaign To Pass Comprehensive Climate Legislation in the United States Failed*: Prepared for the Symposium on: The Politics of America's Fight Against Global Warming, Co-sponsored by the Columbia School of Journalism and the Scholars Support Network, Report Commissioned by Lee Wasserman at the Rockefeller Family Fund in conjunction with Nick Lehman, Columbia School of Journalism. P. 39.

for deep reductions in GHG emissions, specific allowance allocations, large-scale use of offsets and other cost-containment measures, and CPs.[24] Frequently, the best Washington-based coalitions can achieve is to agree on principles to guide the development of legislation. Although this can be a useful input into the legislative process, such principles are often too general to make much of an impact on the actual legislation. Some would argue that the USCAP Blueprint suffered from the opposite problem; it was too detailed. At the end of the day, members of Congress have to exercise their own judgment regarding the substantive elements to be included in legislation and the impacts it will have on their districts and states. And the appearance of incorporating too much of an outside group's proposal in legislation can provide opponents with direct lines of attack.

The second issue that surprised me was the level of deference that supportive members of Congress granted to USCAP. Given its membership, USCAP was, in the annals of Washington policy-making, an historic coalition. However, it only included 30 members, and although diverse, its members were big companies and prominent environmental groups. In addition, and similar to the views expressed by some USCAP members, the coalition's close affiliation with legislators sympathetic to the cause of passing legislation threatened to cast the coalition as a partisan group.

From Natsource's perspective, the global carbon markets were in decline and our funds had completed most of their contracting for CERs. We were increasing efforts to grow the business in the US. Legislation would create the largest carbon market in the world and the company was perfectly situated to capitalize on this opportunity based on the track record we had established contracting for CERs and our long-standing relationships with the large emitting companies in the US.

We were primarily engaged in two activities. The first was working on the climate change bill for a group of utilities that owned and operated coal-fired power plants. Our major focus was providing advice on the market-based mechanisms with an emphasis on the offset and trading

---

[24] United States Climate Action Partnership. *A Blueprint for Legislative Action.* 2009.

provisions. And the second activity was working with a group of electric utilities to design a fund that would source offsets for compliance buyers. This was similar to the process we employed in developing GG-CAP.

American Clean Energy and Security Act of 2009

Three stages in the legislations evolution are worth noting: the discussion draft, the markup draft, and the final legislation that went to vote. During the process of developing the legislation, it became a kitchen sink of policies and grew to an unwieldy size.

The 648-Page Discussion Draft

Following the release of the USCAP Blueprint and subsequent hearing, the legislative process began to play out. On 30 March 2009, Chairmen Waxman and Markey released a 648-page discussion draft of the American Clean Energy and Security Act of 2009 (ACES) of 2009.[25] A summary of the draft distributed by committee Democrats communicates that the global warming provisions were modeled on USCAP's Blueprint.[26] A discussion draft released by the Chairman is an important part of the legislative process. It means that enough work has been done to develop base legislation, but it offers members of Congress, particularly those on the committee, the opportunity to provide input into the process prior to the formal introduction of legislation. Importantly, from a political perspective, it is also a way to determine where members of Congress and the interest groups stand on the issues. It is one thing to discuss and react to concepts that are under development, but legislative language is concrete. One of the missing elements from the discussion draft was detail regarding the

---

[25] Available from: https://wayback.archive-it.org/4949/20141224014808/http://democrats.energycommerce.house.gov/sites/default/files/documents/ACES-2454-Discussion-Draft-2009-3-30.pdf [Accessed 17 December 2015].

[26] See Discussion Draft Summary, Available from: https://wayback.archive-it.org

allocation of allowances/permits. This would be subject to negotiation and used to secure votes for passage.

Prior to the formal introduction of ACES, 14 hearings were held in the Energy and Environment Subcommittee and the full Energy and Commerce Committee on climate and energy-related issues; this included four days devoted to the discussion draft of the legislation.[27] Nearly 70 witnesses testified on the legislation and related issues. Hearings also play an important educational role in the legislative process. Devoted to specific issues, they provide committee members the opportunity to ask questions of expert witnesses on topics of interest and to gain a greater understanding of the legislation that they will ultimately vote on and determine how it could affect their constituents.

Key Provisions of HR 2454: The 932-Page Markup Draft

Chairmen Waxman and Markey formally introduced HR 2454 on 15 May 2009. This 932-page bill was ready to be marked up by the committee.[28] In a committee markup, members of the committee are provided the opportunity to offer and vote on amendments to the bill and its final passage out of committee. These votes generally take place prior to the full House of Representatives voting on a bill. In addition to the climate change components of the legislation, HR 2454 included, among others, numerous CPs designed to achieve a multitude of objectives including increased use of renewables, improving energy efficiency, and development/deployment of low- and non-emitting technologies to name a few.

The messages that Democrats attempted to communicate at the introduction of the legislation were notable. A press release communicated, '[t]he legislation would will create millions of new clean energy jobs, save consumers hundreds of billions of dollars in energy costs, promote America's energy independence and security, and cut global warming pollution', said Chairman Waxman.[29] In recognition of the current economic situa-

---

[27] The list of hearings is available from: https://wayback.archive-it.org
[28] This version of the legislation is available from: https://wayback.archive-it.org
[29] This statement is available from: https://wayback.archive-it.org

tion, the legislation was marketed primarily as a job-creation measure. It was breathtaking in its ambition and complexity, which will be described later. If enacted, it would remake the nation's energy infrastructure.

The effort to pass the legislation raised some important questions. A few of the pivotal ones included (i) was the public ready for such far-reaching change, (ii) could the legislation garner sufficient public support at the same time health care reform was being debated, and (iii) would Chairmen Waxman and Markey be able to manage the legislative process? The answers to these questions would be complicated by an economy still in decline and increasing Republican attacks on the bill.

A brief description of four important sections included in the nearly 1000 pages of legislation follows. These include the global warming title, carbon and energy market regulation, and CPs. There were hundreds of other programs included in the legislation designed to address energy and environmental issues and other objectives.

Global Warming Title

1. *Cap-and-trade:* The legislation would establish an economy-wide GHG cap-and-trade program, covering approximately 85% of US emissions, and impose a declining cap. The bill would require modest reductions in GHG emissions beginning in 2012 from 2005 levels, decline to 17% below 2005 levels in 2020, and ultimately require reductions of over 80% by 2050.[30]

2. *Allowance allocation:* The allocation of allowances may be the most contentious issue in establishing a cap-and-trade program. This is because they have significant monetary and political value. The allocations included in ACES attempted to achieve many different substantive and political objectives. These included:

   • **Consumer protection.** A climate change program has the ability to increase the costs of electricity in regions of the countries dependent on fossil fuels, particularly coal, to meet their energy requirements. As

---

[30] The reduction requirements can be found in Title VII, Global Warming Pollution Reduction Program and Targets, Part A, Available from: https://wayback.archive-it.org

such, nearly a third of the program's allowances were allocated to electric distribution companies regulated by states. The allocation to electric distribution companies was designed to cushion the legislation's impacts on consumers in the mid-west and southeast, the regions in the US most dependent on coal. This was vital to winning the support of the Edison Electric Institute, the trade association representing the interests of investor-owned utilities. Approximately 10% of allowances were allocated to gas distribution companies subjected to state regulation to mitigate price increases caused by the legislation. Out of recognition that EU companies benefitted from their allowance allocations in the EU ETS, the electric and gas electric distribution companies receiving these allocations were required to use them to protect consumers from increases in electricity and gas prices. A smaller allocation was provided to states that used home heating oil and propane for the same purpose.[31]

- **Protection of lower- and middle-income American's from higher energy prices.** A climate change program has the ability to impose disproportionate impacts on low- and middle-income groups because they spend a greater percentage of their incomes on energy than those with higher incomes. To guard against this, 15% of the allowances would be auctioned at a minimum price to shield these groups from increases in energy prices. Proceeds from the allowance sales would be distributed to these income groups in various tax credits and transfer payments.[32] Similar to guarding against regional disparities, no bill that imposed significant impacts on the poor and middle class could become law. Democrats in particular would be hard-pressed to support legislation that disproportionately impacted lower-income groups regardless of their concern with climate change.

- **Protecting energy-intensive and trade-sensitive industries.** A climate change program has the ability to have direct and indirect impacts on energy-intensive, trade-sensitive industries such as iron

---

[31] A summary of these allocations can be found in the consumer protection section of a document titled, 'Proposed Allowance Allocation. Available from: https://wayback.archive-it.org [Accessed 17 December 2015].

[32] This allocation can be found in the consumer protection section. Available from: https://wayback.archive-it.org

and steel, chemicals, aluminum, pulp and paper, and cement and their workers. Direct impacts are caused by the expenditures firms need to make to meet their GHG emissions cap. Indirect impacts result from paying higher prices for energy. The program would also disadvantage these companies and create job loss if their competitors in other countries were not confronted with similar costs to reduce their GHG emissions. In addition, there is the potential that GHG emissions could increase elsewhere if production shifted to jurisdictions that did not regulate similar activities. This is known as leakage. To guard against these outcomes, 15 % of allowances were initially allocated to these industries and small allocations were also provided for worker transition and worker assistance programs.[33] In a similar vein, 2 % of allowances were allocated to domestic oil refiners.[34] These allocations were essential to secure the support of members of Congress that represented these industries and to guard against job loss, which was vital to secure support from labor unions, an important constituency of the Democratic Party.

- **Clean energy**. Nearly 15 % of allowances, valued at $190 billion, were initially allocated to support the development and deployment of carbon capture and storage (CCS) technologies, energy efficiency, and renewable and transportation technologies.[35]
- **Allocations for other purposes**. Remaining allowances were allocated for domestic and international adaptation and to prevent deforestation.[36]

The allowance allocations in the legislation subjected the legislation to ridicule and attacks by its opponents, which is described later in the chapter.

---

[33] These allocations can be found in the Transition Assistance for Industry and Other Public Purposes Sections. Available from: https://wayback.archive-it.org

[34] These allocations can be found in the Transition Assistance for Industry section. Available from: https://wayback.archive-it.org

[35] These allocations can be found in the Clean Energy and Energy Technology Section. Available from: https://wayback.archive-it.org

[36] See Proposed Allowance Allocation. Available from: https://wayback.archive-it.org

3. *Cost containment and mitigation of allowance price volatility:* Another concern with cap-and-trade programs is the potential for high allowance prices and price volatility. The legislation included several measures designed to control covered sources costs to comply with GHG emissions reduction requirements and to minimize the potential for price volatility. Among others, these included full trading and unlimited banking of allowances, limited borrowing,[37] the use of a large volume of offsets, a strategic reserve, and an auction:

- **Offsets.** The legislation authorized the annual use of two billion tons of domestic and international offsets per year for compliance.[38] EPA analysis of the bill determined that the use of international offsets would reduce allowance prices by 89 %.[39] For a variety of reasons, those of us experienced with offset programs had strong doubts as to whether such large volumes of domestic and international offsets would materialize. On the domestic side, we did not know if one billion offsets would become available because the cap-and-trade program covered nearly 85 % of national emissions. On the international front, assumptions were made that new types of mechanisms would be developed in the international negotiating process. This was always speculative. Second, the analysis could not account for the competition for international offsets that the US would face from other buying regions attempting to comply with GHG emissions targets. And finally, international allowances could only also be used in the program subject to meeting certain conditions.
- **Strategic Reserve.** Concerned with the potential for allowance price volatility (particularly high prices), the legislation created a strategic allowance reserve that made allowances available to covered

---

[37] The trading, and banking and borrowing provisions can be found in Title VII, Part C. Available from: https://wayback.archive-it.org

[38] The offsets provisions can be found in Title VII, Part D. Available from: https://wayback.archive-it.org

[39] US Environmental Protection Agency. *EPA Analysis of the American Clean Energy and Security Act OF 2009.*

sources when prices reached a certain level. This served to cap compliance costs.[40]

- **Auction.** Concerned with the potential for very low allowance prices that do not provide incentives for investment in environmentally beneficial activities, quarterly auctions were established. The legislation established a minimal price at which allowances could be purchased for in the auction.[41]

Carbon and Energy Market Regulation

Concerned with the creation of a potentially large new commodity market and the potential for speculation and market manipulation, the legislation created a new program to regulate allowances, offsets, and renewable energy credits that had been created by another program. Jurisdiction was split between the Federal Energy Regulatory Commission and the Commodity Futures Trading Commission. Provisions were also included to govern energy commodity derivatives and credit default swaps. If additional legislation was passed that reformed the regulation of derivatives, sections of the legislation included in this title would be repealed.[42]

Complementary Policies: Clean Energy and Energy Efficiency

ACES included many CPs, resembling the approach of the EU Climate and Energy Package and the State of California's climate change program. Further, it was consistent with USCAP's Blueprint and supported by environmental organizations. Its ambition is still surprising to me six years later.

Similar to the analysis of the EU ETS performance in Chap. 4, I do not believe that policy-makers fully considered the impacts on the carbon market that would be caused by interaction with the CPs, nor did they

---

[40] The Strategic Reserve can be found in Title VII, Part C. Available from: https://wayback.archive-it.org

[41] The Auction can be found in Title VII, Part H. Available from: https://wayback.archive-it.org

[42] The provisions creating the regulatory program can be found in Title VIII, Additional Greenhouse Gas Standards, Part IV. Available from: https://wayback.archive-it.org

anticipate the political backlash that occurred. A brief description of some of these programs follows:

- **Clean Energy**. Title I of the legislation was devoted to the development and deployment of non-emitting technologies. It included a combined efficiency and renewable electricity standard requiring retail electric suppliers to meet 20 % of their demand with renewable sources of energy, a portion of which could be met by increasing energy efficiency.[43] This was similar to the renewable portfolio standards in place in more than 30 US states and the EU renewable energy directive, although EU member states were responsible for carrying out the policies for achieving it. It also included programs to stimulate development and deployment of low- and non-emitting technologies including CCS,[44] electric vehicles,[45] smart grid,[46] and others.

- **Energy Efficiency**. Title II of the legislation included many programs to improve energy efficiency. Programs were developed to improve efficiency in residential and commercial buildings through new codes and labeling programs, and to improve the performance of a myriad of energy-using products including lighting and appliances.[47] Provisions in the energy efficiency title were also included to improve efficiency in the industrial and transportation sectors and the federal government.

- The programs in Titles I and II would accomplish their objectives using several policy instruments including allowance allocations, direct spending, research and development, mandates including product standards, incentive programs, and subsidies. Many federal agencies would be tasked to implement these programs.

---

[43] The provisions creating the program can be found in Title I, Clean Energy, Subtitle A. Available from: https://wayback.archive-it.org

[44] The CCS program can be found in Title I, Subtitle B. Available from: https://wayback.archive-it.org

[45] The clean transportation program can be found in Title I, Subtitle C. Available from: https://wayback.archive-it.org

[46] The smart grid program can be found in Title I, Subtitle E. Available from: https://wayback.archive-it.org

[47] These programs can be found in Title II, Energy Efficiency, Subtitles A and B. Available from: https://wayback.archive-it.org

The Legislative Process Plays Out

This section describes the supporters' and opponents' messages and strategies in their attempt to achieve their objectives and the outcomes of the committee markup and the house debate.

Democratic Efforts and Messages Prior to MarkUp

The Democrats' efforts in the period prior to markup was to negotiate agreements on several important issues such as the clean energy standard, allowance allocations, and offset use necessary to secure votes for passage in the committee with an eye to the coming debate in the full house. Importantly, several members of the committee perceived to be more moderate than Congressmen Waxman and Markey were at the center of these negotiations. Congressman Rick Boucher was pivotal in negotiating the allowance allocations for electric distribution companies and offset use. Congressman Mike Doyle (D-PA), whose congressional district included Pittsburgh, was a key participant in the negotiation of the allocation for energy-intensive and trade-sensitive industries. Although it was natural that these members were involved in negotiations on issues vital to their constituents, their engagement was highly publicized in an attempt to provide some comfort to like-minded Democrats in both the committee and the full house.

The Democrats also took great pains to attempt to communicate the legislation's impact and benefits. Prior to markup, they relied primarily on an April 2009 analysis of the discussion draft of the legislation undertaken by the EPA. The analysis concluded that the legislation would cost 27–38 cents per day and allowance prices would range between $13 and $22 in 2015 and 2020. It also indicated that the bill would have little impact on economic growth and that electricity from zero or low-carbon sources would double by 2030 compared to a no-policy scenario.[48]

---

[48] See the summary of EPAs Preliminary Economic Analysis of the discussions draft. Available from: https://wayback.archive-it.org

The Republican Strategy

The large majority of Republicans had expressed opposition to any actions to reduce GHG emissions for many years. It was clear from the hearings and tone of the debate that legislation would only pass in the committee with Democratic votes. With one exception, the Republicans were not going to play. Their views on the bill, key messages, specific lines of attack, and strategies they would use in an attempt to derail it in committee and defeat it were foreshadowed in two pieces of testimony presented by their allies in the committee hearings on the legislation.

In his testimony, Newt Gingrich, the former Republican Speaker of the House, argued that the bill was bad for national security and the economy. It was a tour de force of economic claims that Republicans have made against Democrats for the last several decades. He indicated the bill represented a $650 billion to $1.9 trillion tax on energy.[49] This was in contrast to the Vietnam War, which cost $700 billion; the New Deal, which cost $500 billion; and the National Aeronautics and Space Administration, which cost $850 billion since its creation.[50] The bill would dramatically increase the price of gasoline, electricity, home heating oil, and natural gas.[51] He cited analysis of other cap-and-trade legislation that would cause a massive loss of jobs.[52] It did not matter that the job losses he cited was from analysis of other proposals. And he concluded by denouncing the legislation for being overly bureaucratic, encompassing a command-and-control philosophy, and by citing a provision in the legislation that authorized the Secretary of Energy to set standards for Jacuzzis.[53]

Gingrich cited other reasons for opposing the legislation. These included that the US would be disadvantaged by the failure of China and India to enact policies that would reduce GHG emissions, and the similarities between the failed housing market that had a hand in bringing

---

[49] N. Gingrich. *Statement before the House Subcommittee on Energy and the Environment on HR 2454.* Available from: https://wayback.archive-it.org P. 4.

[50] Gingrich, N. *Statement before the House on HR2454.* P. 5.

[51] Gingrich, N. *Statement before the House on HR2454.* P. 5.

[52] Gingrich, N. *Statement before the House on HR2454.* P. 7.

[53] Gingrich, N. *Statement before the House on HR2454.* P. 7.

down the US economy, and a \$1–2 trillion carbon market that the legislation would create.[54]

Another piece of testimony on the legislation was delivered by Myron Ebell of the Competitive Enterprise Institute, a group that describes itself as focusing on regulatory issues 'from a free market and limited government perspective'.[55] The group has been aligned with Republicans' views on the issue of climate change for many years. Echoing one of former Speaker Gingrich's arguments, Ebell argued that cap-and-trade would be the largest intervention in US citizen's lives since World War II,[56] equating limits on GHG emissions to rationing important consumer goods such as gasoline and food products.[57] He argued that while opponent's opposition to ACES is primarily based on cost, this intervention is a significant intrusion into citizen's economic and economic freedoms. He also called cap-and-trade an indirect tax and said, '[i]f Title III or something like it were enacted, it would probably be the biggest tax increase in the history of the world'.[58] Finally, Ebell described the EU ETS failure to reduce emissions and its success in increasing electricity rates and transferring wealth to power companies. He indicated that the wealth transfers to power companies (presumably in the form of allowance allocations) were a reason why many corporate members of USCAP supported the bill.[59] Gingrich's testimony also implied that business was supporting the legislation to get their cut of the \$2 trillion that the legislation would raise. He characterized these companies as panting dogs, 'vying for their cut of the green spoils'.[60]

In contrast to former House Speaker Gingrich's and Ebell's testimony, the testimony of the legislation's most prominent supporters, former Vice President Al Gore and former Republican Senator John Warner (R-VA), a respected Republican on national security issues

---

[54] Gingrich, N. *Statement before the House on HR2454.* PP. 6–8.

[55] Ebell, M. *Testimony before the US House of Representatives Committee on Energy and Commerce.* Available from: https://wayback.archive-it.org P. 1.

[56] Ebell, M. *Testimony before the Committee on Energy and Commerce.* P. 2.

[57] Ebell, M. *Testimony before the Committee on Energy and Commerce.* P. 2.

[58] Ebell, M. *Testimony before the Committee on Energy and Commerce.* P. 2.

[59] Ebell, M. *Testimony before the Committee on Energy and Commerce.* P. 3.

[60] Gingrich, N. *Statement before the House on HR2454.* P. 7.

who had previously authored an economy-wide GHG cap-and-trade proposal, did not focus on the economic impacts of the legislation.[61] Whereas it appeared the Republicans closely coordinated their messages in expressing their opposition to the legislation, it does not appear supporters did.

The apparent difference in coordination may not seem important. However, like all political issues, legislation becomes associated with a narrative. The overwhelming majority of the public does not have the time to sift through the arguments for and against legislation, particularly on an issue as complex as climate change, and come to their own conclusions. They will throw their support to the side that echoes its views and communicates them effectively. This is particularly at a time when the US economy had come close to falling into a depression and was losing hundreds of thousands of jobs per month. It is safe to assume that opponent's predictions of the legislation's adverse economic impacts were more effective in swaying public opinion than Vice President Gore's admonition to address the risks of climate change.

Committee Markup

The bill was marked up by the full Energy and Commerce Committee during 21–24 May 2009. The committee considered 96 amendments and adopted 36 and the bill passed by a 33–25 vote.[62] Thirty-two Democrats, or approximately 90 % of those serving on the committee, voted in favor of the legislation and four opposed it.[63] The Democrats that opposed the bill were members of the Blue Dog Coalition, which is comprised of members that are generally viewed as more conservative than the national party.[64] Two also represented energy-producing states. Congresswoman

---

[61] Former Vice President Gore's and former Senator Warner's testimony is available from: https://wayback.archive-it.org

[62] A summary of committee actions is available from: https://wayback.archive-it.org

[63] The committee vote can be found on P. 356 of the committee report which is available from: https://wayback.archive-it.org

[64] A description of the Blue Dog Coalition is available from: http://bluedogcaucus-schrader.house.gov/members

Bono Mack of California was the only one of the 22 Republicans on the committee that supported it.

This was only the second time a federal climate change bill won the support of a major congressional committee. The first time was in the Senate Committee on Environment and Public Works in the proceeding Congress. However, no one was under any illusion that the prior effort would become law. That was practice for this Congress. The next vote on ACES would be by the full House of Representatives.

In preparation for the house debate, the Democrats relied on an updated EPA analysis of the bill passed by the committee and a new one provided by the non-partisan Congressional Budget Office (CBO).[65,66] According to a committee summary of the EPA analysis, the committee bill would cost the average household $80–111 per year or equivalent to 22–30 cents per day. Allowance prices would be $13 in 2015 and $16 in 2020, and economic growth would not be damaged.[67] The CBO analysis estimated that the net economy-wide costs of the legislation would be $22 billion or approximately $175 per household in 2020.[68] Importantly, the analysis concluded that in 2020, the lowest-quintile-income group would see a benefit of $40 per year, the group in the middle would see a net cost of $235 per year, and the highest-income group would see a cost from the legislation of $245 per year.[69] According to CBO, households would receive $28 billion in allowance value or 30 %, business would receive $47 billion or 50 %, 10 % would be provided to state and federal government to spend and 7 % would be spent overseas on adaptation, technology, and to avoid deforestation.[70]

---

[65] Environmental Protection Agency. *EPA Analysis of the American Clean Energy and Security Act.* 2009.

[66] Congressional Budget Office. *The Estimated Costs to Households From the Cap-and-Trade Provisions of H.R. 2454.* 2009.

[67] See the summary of EPAs Economic Analysis of the 'American Clean Energy and Security Act of 2009'. Available from: https://wayback.archive-it.org

[68] Congressional Budget Office. *The Estimated Costs to Households.* P. 2.

[69] Congressional Budget Office. *The Estimated Costs to Households.* P. 2.

[70] Congressional Budget Office. *The Estimated Costs to Households.* PP. 5–6.

The House Debates Final Passage of the 1400-Page Bill

Debates on the floor on legislation in the US House of Representatives and Senate are cataloged in the Congressional Record. Members of Congress are provided time to communicate their views on legislation prior to a final vote being held. A review of the debate on ACES indicates how partisan and rancorous the climate issue had become.[71]

The legislation debated by the house had grown to over 1400 pages. One of the new provisions created a separate offset program for agricultural and forestry-related activities.[72] In many ways, the program was similar to the offsets provision included in the bill's Global Warming title. The inclusion of a separate agricultural title in the bill was about Washington politics, committee jurisdiction, and power. The chairman of the House Committee on Agriculture, Congressman Collin Peterson (D-MN), wanted the Department of Agriculture (DOA), not the EPA, to govern offsets that would be created by agricultural and forestry activities. The agricultural community creating the offsets preferred to be regulated by DOA, which was known for being an advocate for farm interests. And if the program was to be regulated by DOA, the congressional agriculture committees gained jurisdiction over a new program based on its oversight of the Department. Ordinarily, a companion program of offsets may not have been developed. However, Peterson's vote in support of the legislation was at risk as were some of the Democratic members that served on the committee. Given that the vote was going to be close, there was no margin for error. Thus, to gain the necessary votes, the program was created.

The theme of Democrats' messages in support of the bill remained the same. They argued that passage of the bill was essential to break the nation's dependence on foreign oil, would create millions of clean energy jobs, and address global warming. And, they argued that the cost to accomplish these far-reaching objectives would be affordable. In

---

[71] To review the debate on the house floor on the American Clean Energy and Security Act, see the Congressional Record, 26 June 2009, 155 (98) H 7619–7687, Available from: https://www.congress.gov/crec/2009/06/26/CREC-2009-06-26-house.pdf [Accessed 18 December 2015].

[72] The provision to create an Offset Credit Program From Domestic Agricultural and Forestry Sources can be Found in Title V, Agricultural and Forestry Related Offsets Program, Subtitles A and B. Available from: https://wayback.archive-it.org

Chairman Waxman's floor statement, he referenced the CBO analysis which concluded that the legislation would cost households less than 50 cents per day and the EPA analysis which put the cost at 22–30 cents a day.[73] In addition to these arguments, both Congressman Dingell and Boucher highly respected members of the committee that were key to forging important legislative compromises prior to and during markup raised the bogeyman of EPA regulation of GHGs to urge their colleagues' support of the bill. During this time period, EPA was developing its endangerment finding to determine if GHGs endangered public health and welfare as required by the Massachusetts versus EPA Supreme Court decision. An affirmative determination would require EPA to regulate GHGs. Congressmen Dingell and Boucher argued that a legislative solution created by the Congress would be more efficient and economically sustainable to address climate change than would burdensome regulations developed by EPA.[74]

The Republicans were strident in their opposition to ACES during the floor debate. Their messages regarding the bill were consistent with those communicated by former Speaker Gingrich during his committee testimony. First and foremost, the legislation was consistently characterized as a national energy tax, cap-and-tax, and a job-destroying energy tax, which would wreak havoc on the national economy. The Republicans argued the legislation would be the largest tax increase in history, quoting several studies that concluded that passage would result in millions of lost jobs and increase the prices for energy including electricity, gasoline, and other energy commodities.

Because all politics is local, Republican house members from Florida, Iowa, Michigan, Missouri, Pennsylvania, Nebraska, and Tennessee all cited studies done by their local utilities and constituent businesses detailing the adverse impacts the legislation would have in their congressional districts. Another line of economic attack was that China and India would not follow the US in limiting their GHG emissions, which

---

[73] Waxman, Representative H. Floor statement on ACES. *Congressional Record.* 2009: H 7619.

[74] To review the Representatives Boucher's and Dingell's statement on ACES, see pages H 7623 and H 7625 in the Congressional Record.

would result in a loss of manufacturing jobs. All of these messages were amplified by television ads.

The attacks did not stop there. Another message that resonated was that the legislation was heavy-handed. The programs included in the bill would require a massive new bureaucracy (many of the programs were in the clean energy and energy efficiency titles cited previously) centralized in the nation's capital and represent a federal takeover of national energy policy. Of course, this bureaucracy would intrude on citizen's economic freedoms and liberties. The creation of new agencies was ridiculed, as were the mandates and rulemakings that would be necessary to implement ACES. One member of Congress cited a Chamber of Commerce study indicating that the legislation required 397 new rulemakings.[75] At any time, these arguments resonate with a large portion of US citizens. They were more powerful at the time of the debate given the then-current economic climate and the rise of the tea party in the US, which became an important part of the Republicans electoral base. And for good measure, one additional line of argument was that the legislation would create a large carbon market that would enrich Wall Street at the expense of the public. Opponents even attacked supporter's arguments that ACES would create green jobs. To debunk this message, opponents cited studies that concluded that Spain lost two conventional jobs for every green job it created and that each one cost in excess of $750,000. These arguments were made even though little comparison could be made between the US's and Spain's economies or policies.[76]

Some not familiar with the US policy-making process will wonder why these debates mattered. It should be kept in mind that the senate had not yet considered climate change legislation. The house debate was similar to preparing the battlefield for the upcoming senate debate. It would condition the public's and senators' views on the legislation and serve as a preview for the upcoming senate deliberations.

The bill passed the house by the narrow margin of 219–212 with 211 Democrats and eight Republicans voting for passage. Forty-three, or less

[75] Burton, Representative D. Floor statement on ACES. *Congressional Record.* H 7666.
[76] Burton, Representative D. H 7666.

than 20 %, of the house Democrats voted against the bill while 169 or approximately 95 % of house Republicans voted against.[77]

The Bill Officially Dies in the Senate

The hopes that were raised by the 2008 election that the US would adopt a federal GHG emission reduction program to address climate change were not realized. The USA Senate never took up comprehensive climate change legislation. Within the context of this book, I do not believe it would add much to provide a chronology of the senate's attempt to cobble together climate change legislation. The Waxman-Markey bill, or similar legislation, never had a chance to pass the senate for the reasons that follow.

Procedural Issues

The rules that govern the 100-member, freewheeling senate provides opponents of legislation with significantly more opportunities for obstruction than the rules of the 435-member House of Representatives, which is necessarily governed by more formal rules. The house requires a simple majority of those voting on a bill for passage. With few exceptions, Senators are provided the right to filibuster a bill to block or delay legislative action. Sixty votes are needed to break a filibuster and thus are also required to pass a bill.

Disbursed Power

In contrast to the House of Representatives in which states representation is based on its percentage of the US population, each state has two

---

[77] govtrack.us. *Vote on HR2454*. Available from: https://www.govtrack.us/congress/votes/111-2009/h477 [Accessed 18 December 2015].

Senators. The two Senators from Wyoming, the least populous state, have as much influence and as many votes on legislation as the two Senators from California, the most populous state. In contrast, there were 53 house members from California in 2009, including 34 Democrats, and only one from Wyoming. Because of California's long tradition of supporting environmental measures, this dynamic alone provided a better opportunity for assembling a majority coalition in support of climate change legislation in the house than in the senate.

Regional Composition of the Senate

Another factor that made the passage of climate change legislation in the senate so challenging was its regional composition by party. If one looked at the senate solely based on its membership by party in 2009–2010, they may have thought that there was a good chance to pass a climate change bill. Democrats controlled between 55 and 58 seats during these years, Republicans' membership fluctuated between 39 and 42 seats, and there were two independents that caucused with the Democrats on many issues. Because Democrats were predisposed to support climate change legislation, the numbers made it appear as if only a few Republicans would be needed to pass a bill. However, taking a closer look at the states represented by the Democrats suggests a different line of thinking. Approximately 20 Democratic senators represented energy-producing states, and others represented states with significant manufacturing and agricultural interests which were concerned with the impacts of a climate bill. Some were more predisposed to support climate legislation than others. However, because all of these states could have been greatly affected by such legislation, there was a far greater chance for Democratic defections. This means that far more Republican votes would be required to pass a bill.

All of these dynamics conspired to make it nearly impossible to pass legislation in the senate. And, this does not consider the near-unanimous Republican opposition that existed to climate change legislation in the senate. Senator John McCain, the party's 2008 Presidential nominee who had previously led the charge on cap-and-trade, labeled Waxman-Markey a '1400 page monstrosity'.[78]

There was no way the Senate Environment and Public Works Committee could pass a bill that could secure the 60 votes need in the full senate. Democrats on the committee, which was chaired by Senator Barbara Boxer (D-CA) in 2009, were more liberal than their colleagues in the full senate. The committee passed a bill by a margin of 11–1 in September 2009.[79] One Democrat voted in opposition. Seven Republican members of the committee boycotted the vote.

Following this, Senators John Kerry (D-MA), Joe Lieberman (I-CT), and Lindsay Graham (R-SC) attempted to craft a bipartisan compromise. For many reasons, they could never agree to a bill and the effort to pass legislation officially died in 2010.[80,81] Many viewed this as the official death of climate change legislation in the US.

In the 2010 elections, the Republicans picked up 63 seats, taking control of the House of Representatives. And with that, the prospects for climate change legislation effectively died. That represented the end of climate change policy 1.0 in the US.

Why the Legislation Failed

Although no single issue was responsible for the failure of climate change legislation in 2009–2010, several contributed to it. What follows is a description of some of the dynamics that have been cited as causes for the failure of climate change legislation in the US.

---

[78] Lizza, R (2010). 'As The World Burns: How the Senate and White House missed their best chance to deal with climate change'. *The New Yorker*, 70–83.

[79] This vote is available from: http://www.c2es.org/federal/congress/111 [Accessed 18 December 2015].

[80] Lizza, R. (2010) 'As The World Burns'. This article provides a detailed account of the events which contributed to climate change legislations demise in the US Senate.

[81] Pooley, E (2010) *The Climate War: True Believers, Power Brokers, And The Fight To Save The Earth.* (New York: Hyperion). This book provides a detailed account of the efforts to pass climate change legislation in 2009–2010.

The Economic Landscape

The US economy was in freefall when the climate change debate began in earnest. In the fourth quarter of 2008, growth in the US economy fell by over 6%[82] from the prior quarter and 5.5% in the first quarter of 2009.[83] As mentioned previously, seven million jobs were lost in the last four months of 2008 and in 2009. The stock market had fallen precipitously through the first quarter of 2009. People were frightened during this time. Developing climate change legislation is difficult in good economic times. It was much more difficult in a time of plunging economic performance.

The deep recession that took hold of the economy made it easier for the opponents to scare the American public about potential adverse economic impacts climate change legislation would have. And as would be expected, restoring the economy was a far higher priority of the US public than environmental issues when the debate commenced. A New York Times CBS News Poll in January 2009 showed that 58% of the public believed stimulating the economy should be the priority, while 33% supported initiatives to protect the environment. A similar poll in April 2007 showed nearly the opposite. Over 50% of the public thought protecting the environment should be emphasized, while approximately 35% believed that stimulating the economy should be a higher priority.[84]

It is extremely difficult, if not impossible, to assess the impact of one factor in killing the climate change bill. One analyst looked at the unemployment rates when important US environmental laws were passed and the rates when climate change legislation was being debated. It concluded that the major environmental statutes such as the CAA of 1970, The Clean Water Act of 1972, the Endangered Species Act of 1973, and the Resource Conservation and Recovery Act of 1976 all passed when the unemployment rate was lower than 6%. Six major environmental laws

---

[82] US Department of Commerce. *Gross Domestic Product: 4th Quarter 2008 (final)*. [Press Release] 27 February 2009.

[83] US Department of Commerce. Gross Domestic Product, 1st Quarter 2009 (final). [Press Release] 25 June 2009.

[84] Layzer, J. (2011) 'Cold Front' in Skocpol, T., L. Jacobs (eds.) *Reaching for a New Deal*. (New York, NY. Russell Sage Foundation). PP. 321–385.

were passed when unemployment was over 7% and none passed when the unemployment rate was higher than 7.7%.[85]

Unemployment was skyrocketing during the time when climate change legislation was being debated. It was already at 7.8% when President Obama took office, climbed to nearly 9% by the end of March 2009 when the house was developing legislation, increased to 9.5% when it passed the full house, and reached 10% in October 2009 when the senate was in its initial phases of considering it. It dropped minimally to 9.4% by October 2010 when the legislation was pulled in the senate.[86] Unemployment was much higher during the climate change debate in 2009–2010 than when other, less ambitious environmental legislation was passed.

Increased Partisanship and Polarization in Washington, DC

It is no secret that partisanship has increased in the US political system and become more polarized making it extremely difficult to solve problems. This contributed to the difficulty of passing climate change legislation in 2009–2010.

My Personal Views on Partisanship

Political scientists use many types of data to estimate the level of partisanship in the US political system. I will borrow some to illustrate the growing chasm between Democrats and Republicans on environmental issues and climate change. Before doing so, I wanted to provide my views on this topic based on my 30 years of experience in Washington, DC.

---

[85] D. Weiss. Anatomy of a Senate Climate Bill Death. This material [article] was created by the Center for American Progress. Available from: https://www.americanprogress.org/issues/green/news/2010/10/12/8569/anatomy-of-a-senate-climate-bill-death/[Accessed 18 December 2015]. PP. 3–4.

[86] US Department of Labor. *Labor Force Statistics from the Current Population Survey.* Available from: http://data.bls.gov/timeseries/LNS14000000 [Accessed 18 December 2015].

I moved to Washington to study politics in graduate school. The common thread of my career has been engagement in energy and environmental policy issues and markets as a lobbyist, consultant, facilitator, government official, and in managing a business. I have worked closely with representatives from industry, environmental and consumer groups, the research community, and government officials. During my time in Washington, I served as a facilitator for the Keystone Center, a non-profit institution that brings leaders from interest groups together to discuss environmental and energy issues with the goals of achieving consensus. It was in this position that I learned how pivotal dialogue, the creation of trust, and building relationships was to the public policy process. Using this model, Keystone participated in the debate that led to far-reaching change in the electric power industry in the 1980s and 1990s. I was a lobbyist during debates of the CAAA of 1990 (the last comprehensive overhaul of the CAA) and the EPACT of 1992 before working in the Executive Branch of the US government in the first Clinton Administration from 1993 to 1996. I continued to work on energy and environmental issues, primarily climate, from 1996 to 2000, as a consultant prior to joining Natsource.

The debates over the CAAA and EPACT legislation, both of which were comprehensive in nature, were the polar opposites of the climate change debate. They were contentious and partisan, as they should have been, because members of Congress advocated for policies to protect their constituents and create new opportunities. But these debates were driven more by proposals' impacts on regional economies. And although the regional impacts of climate change legislation were thoroughly discussed as indicated above, partisanship played a much larger role in the 2009–2010 debate than in prior ones. There appeared to be little, if any, cooperation between the parties. Republicans generally expressed their opposition to climate change legislation when the process was initiated and it never wavered.

In the house, passage of the CAAA of 1990 and EPACT of 1992 was achieved because of the establishment of bipartisan coalitions in the Energy and Commerce Committee. The committee had a center. A group of moderate and conservative Democrats including Congressmen Billy Tauzin, Jim Slattery (D-KS), Jim Cooper (D-TN), and Rick Boucher and others were able to work with committee Republicans to fashion effective and balanced agreements. The committee leadership seemed to encourage this. This also contributed to cooperative relationships between the committees' professional staff. And, to illustrate the level of bipartisanship, the CAAA of 1990 passed the house by a vote of 401 to 25. One hundred and fifty-three of the voting 173 Republicans, or approximately 90%, voted in support.[87] The senate vote in support was 89 to 10. Thirty-nine of the 44 voting Republicans, or approximately 90%, voted in support.[88] A similar process characterized the debate over EPACT and it also passed the house and senate with large bipartisan majorities.

The circumstances regarding these debates were not entirely analogous to the climate change debate of 2009–2010. For one, George H.W. Bush had been president during the CAAA and EPACT debates; he needed to work with Democratic majorities in the house and senate to get anything done. Also, Republicans were more willing to cooperate with a president from their own party. However, securing a 90% Republican majority in support of CAA legislation, which regulated many important US industries, was an exceptional feat at any time, regardless of who the president was. It was a different time. This is clearly not possible today for many reasons including the prominence of the tea party within the Republican Party which views compromise with disdain.

My view is that partisanship increased dramatically following the 1994 elections when I was working in DOE. This was when the Republicans took control of the House of Representatives for the first time in four decades, following President Clinton's attempt to pass a BTU tax, reform the health care system, and pass a controversial crime bill. The Republicans ran on a platform called the Contract with America.[89] Although many of the items appeared to be common sense, the Republicans moved to make sharp cuts in spending, including environmental programs and to weaken important environmental laws. Some also proposed to terminate entire

---

[87] govtrack.us. Vote on S. 1630. Available from: https://www.govtrack.us/congress/votes/101-1990/h525 [Accessed 18 December 2015].

[88] govtrack.us. Vote on S. 1630. Available from: https://www.govtrack.us/congress/votes/101-1990/s324 [Accessed 18 December 2015].

[89] The Contract with America is available from: http://www.nationalcenter.org/ContractwithAmerica.html [Accessed 18 December 2015].

federal agencies including the DOE. This resulted in two shutdowns of the federal government in 1995 and the beginning of 1996. And although political scientists may provide data indicating that the level of partisanship remained consistent, I disagree. The tone of the debate changed. It was less civil and respectful between the parties than in prior times.

With regard to climate change, partisanship has been the rule since the issue became prominent in the early 1990s. There has been little, if any, cooperation between the parties in fashioning a response. The differences have existed in domestic policy and in the international negotiations. The polarization was evident in two business coalitions comprised of conservative members: the Global Climate Coalition and the Climate Council, which generally opposed any and all efforts to address climate change.

For the most part, Democrats believe that human activity causes climate change and that actions are required to reduce GHG emissions—Republicans, in general, not so much. The debate over the BTU tax and KP during the Clinton Administration can only be characterized as rancorous. In the Bush years from 2001 to 2008, the Administration generally refused to participate in the climate change debate. Let's focus on what happened in the run-up to the debate on ACES.

2006 to 2010

There is a significant body of literature detailing the partisan differences between Democrats and Republicans. Although I will provide some historical context, my focus will primarily be on the 2006–2010 time frame given the effort made to pass ACES in 2009–2010. One analyst comprehensively details increasing partisanship and some of the reasons for it, in a postmortem of the 2009–2010 debate on climate change legislation.[90] A partisan gap of 24 points between congressional Democrats and Republicans, which existed on environmental issues in 1970, and 29 points in 1990, spiked to 63.5 in 2000 and 73.5 in 2010.[91]

---

[90] For an excellent description of the partisan nature of the climate change debate, see Skocpol, T. Naming the Problem: What It Will Take to Counter Extremism and Engage Americans in the Fight Against Global Warming PP. 55–95.

[91] Skocpol, T. Naming the Problem. PP. 60–61.

The analysis details that while Democratic officials' views remained in line with their voters on environmental issues, Republican officials' views on the environment began to diverge with their supporters in the 1990s, which is consistent with the time climate change became a prominent issue in the US. Research cites the publication of 141 anti-environment books developed in the period of 1972–2005. Many focused on increasing doubts and raising questions regarding climate science. One hundred and thirty of these publications had ties to conservative think tanks, which are frequently funded by wealthy conservatives.[92] Although nothing is 100 % conclusive, it appears that the think tanks and their wealthy benefactors, which for the most part represent the elites, implemented a coordinated strategy to influence conservative voters and Republican officials' views on climate change. In addition to the books, the conservatives aggressively used talk radio and other media to raise doubts about climate science, the ideological views of those supportive of taking action and the high costs of reducing GHG emissions.

As mentioned previously, a Gallup poll indicated growing public concern with climate change in 2006 and 2007 consistent with the release of the IPCC's AR 4 and Al Gore's *An Inconvenient Truth*. Other analysis confirmed these trends from 2005 to 2007.[93] Concern with the issue grew among Democrats and Independents, and, to a lesser degree, among Republicans, increasing the potential for government action.[94] The analyst argues that opponents recognized this possibility and moved aggressively to thwart it.[95] And then, public concern with climate change dropped beginning in the middle of 2007 and through 2008.[96]

The researchers that confirmed Gallup's polling result of increased concern with climate change attributed the largest portion of the decline in public concern to partisan debates.[97] They conclude that public opinion is influenced by the way stories are covered through the media. It described how climate change issues were covered by ABC, a mainstream media

---

[92] Skocpol, T. Naming the Problem. P. 67.
[93] Skocpol, T. Naming the Problem. PP. 72–73.
[94] Skocpol, T. Naming the Problem. P. 73.
[95] Skocpol, T. Naming the Problem. P. 74.
[96] Skocpol, T. Naming the Problem. P. 74.
[97] Skocpol, T. Naming the Problem. PP. 74–75.

outlet, and FOX, generally believed by many to be a conservative media outlet, during this period.[98] Both outlets increased coverage of climate change in 2006 and 2007 (consistent with the release of AR 4 and *An Inconvenient Truth*), covered it less in 2008 (consistent with the economic downturn and the presidential campaign), and then increased it again in 2009–2010 when the Waxman-Markey legislation was being debated.[99]

ABC presented the IPCC's AR 4 as definitive science.[100] Much of its coverage communicated that global warming was real and caused by increased GHG emissions resulting from human activity. FOX took the opposite tact. It generally derided climate science and those such as former Vice President Gore that advocated taking action to reduce GHG emissions.[101] These messages were hammered home by like-minded conservative radio talk show hosts, bloggers, columnists, think tanks, and others that were frequently funded by wealthy benefactors, who reached a large portion of the population, according to analysis. Their attempt to shape the views of Republican voters and candidates in the 2007–2008 appeared to work following the uptick of support in the prior 18 months.[102]

The 2008 elections further increased the polarization. In addition to electing Barack Obama as president, the Democrats picked up 21 seats in the house, increasing their majority to 257 members. This followed the 2006 elections in which they had picked up over 30 seats. The Republicans who survived these two elections and newly elected members were among the party's most conservative and those most opposed to taking action to reduce GHG emissions. During this time frame, the tea party, which espoused personal freedom, increased its prominence in the Republican Party making bipartisanship nearly impossible. They were virulently against anything that President Obama was for, including cap-and-trade. And following passage of cap-and-trade in the house, they

---

[98] Skocpol, T. Naming the Problem. P. 77.

[99] Skocpol, T. Naming the Problem. P. 77–79.

[100] Skocpol, T. Naming the Problem. P. 80.

[101] Skocpol, T. Naming the Problem. PP. 80–81.

[102] Skocpol details the impacts that this effort had on Republicans' views on climate change, and its impact on the Republican primaries including Senator McCain who appeared less vocal in support of addressing the issue than he previously had been.

attacked Democrats at town hall meetings in the summer of 2009.[103] Their loudest vitriol was directed at the eight Republican house members who voted for the cap-and-trade legislation. They were labeled 'cap and traitors' and given five days to change their votes or face organized efforts to remove them from office. The tea party sent out an action alert to its members following house passage saying it was time to 'hammer' the senate. Conservative television commentators, including one from FOX, radio commentators, journalists, and publications like 'Human Events' threatened the eight supporters with primary challenges and called on conservatives to remove them from office.[104]

Conservative elements of the Republican Party moved beyond threats by challenging and defeating incumbents, such as Senator Robert Bennett (R-UT), and Congressmen Bob Inglis (R-SC) and Mike Castle (R-DE), who were determined to be insufficiently conservative in 2010 primary elections. Threatened by ideologically driven, well-financed candidates, incumbent Republicans were not going to buck the trend and support action to address climate change.

Let's look at a real-life example of how a concerted conservative campaign, including the role of tea party conservatives and FOX News, played out and influenced the senate debate on climate change legislation at a pivotal time. As described previously, Senator Lindsay Graham, a long-time ally of Senator McCain, committed to work across the aisle with Senator Kerry to craft bipartisan legislation designed to attract the 60 votes necessary for passage. The day after the effort became public, Graham was attacked in a town hall meeting in South Carolina. He was called 'a traitor,' and was accused of 'making a pact with the devil'.[105] Graham communicated to his partners the need to make progress. Someone familiar with the negotiations indicated that Graham said referring to FOX News, '[t]he second they focus on us, it's gonna be

---

[103] For a description of these meetings, see Chap. 46 in Climate Wars by Eric Pooley.

[104] Sheppard, K. *Conservative activists wage war on Republicans who supported climate bill.* Available from: http://grist.org/article/2009-07-02-cap-and-traitors/ [Accessed 18 December 2015].

[105] Lizza, R. (2010). 'As The World Burns: How the Senate and White House missed their best chance to deal with climate change'.

cap-and-tax all the time, and it's gonna be a disaster for me on the air-waves. We have to move this along as quickly as possible.'[106]

Then FOX News really got in on the act. One of the political problems that emerged in the House passed climate change bill was the treatment of the domestic oil industry. Companies would need to hold allowances for the emissions that resulted from the importation and sales of domestic oil. They were not happy with the allowance allocation that they received in ACES. So efforts were being made in the senate negotiations to come up with something more acceptable to the industry. One of the proposed solutions was analogous to a gasoline tax, which is heresy in the Republican Party. When this was leaked to FOX News, it ran an article on its website titled, 'WH [for white house] Opposes Higher Gas Taxes Floated by S.C. GOP Senator Graham in Emerging senate Energy Bill'.[107] The story was updated two times and the word tax was used 34 times.[108] This would not sit well with the tea party or any conservative in South Carolina. An article in a prominent magazine cited additional tea party attacks on Senator Graham and new efforts by a Newt Gingrich group, called American Solutions, targeting Graham because of the reported gas tax bill and urging conservatives to call Graham's office 'and ask him not to introduce new gas taxes'.[109] These campaigns were aggressive, coordinated, and well funded. Between television, radio, op-eds, letters to the editor, and other means of communication, they were relentless. Few Republicans, or politicians of any Party, if any, can or will stand up to this.

Senate Votes in 2009–2010

How has partisanship played out in voting? Congressional Quarterly (CQ), a long-time publication dedicated to reporting on the Congress, developed a methodology to quantify partisanship from 2009 to 2010. As described by one analyst, CQ developed a party unity score 'based

---

[106] Lizza, R. (2010). 'As The World Burns'.

[107] Lizza, R. (2010). 'As The World Burns'.

[108] Lizza, R. (2010). 'As The World Burns'.

[109] Lizza, R. (2010). 'As The World Burns'.

on a proportion of votes that' 'pitted a majority of one party against the other… [reflecting] that each party's position was different and, and a majority of the Senators voted with their own party'.[110] Less than half of the votes were party line in the 101st Congress in 1989–1990. This increased to 72 % in 2009 in the first half of the 111th Congress in 2009 and to 79 % in 2010. The previous high percentage of such votes was 67 % in 1995–1996.[111]

Regarding partisanship's impact on environmental issues, and climate change in particular, legislative solutions cannot be developed by one party. The effort to pass climate change legislation went from extremely difficult to impossible in 2010 when the Republicans gained 63 seats to take back control of the House of Representatives. And to show how conservative the Republicans had become on this issue, an analyst detailed that 19 of 20 Republican senate candidates in 2010 argued that the science underpinning climate change was in question or wrong.[112]

Money

Money in politics was another dynamic that has been pointed to as affecting the effort to develop climate change legislation. Opensecrets.org detailed more than $500 million spent by electric utilities and oil and gas companies between January 2009 and June 2010 to impact the energy bill.[113] This does not include contributions made by other business interests or resources devoted to grass roots efforts. An analysis by the Center for American Progress concluded that oil companies were six of the seven biggest funders of lobbying and that 70 % of oil and coal companies' contributions went to Republicans.[114]

---

[110] Weiss, D. Anatomy of a Senate Climate Bill Death. This material [article] was created by the Center for American Progress. Available from: https://www.americanprogress.org/issues/green/news/2010/10/12/8569/anatomy-of-a-senate-climate-bill-death/   [Accessed on 18 December 2015]. P. 5.

[111] Weiss, D. Anatomy of a Senate Climate Bill Death. PP. 5–6.

[112] Skocpol, T. Naming the Problem. P. 90.

[113] Weiss, D. Anatomy of a Senate Climate Bill Death. P. 7.

[114] Weiss, D. Anatomy of a Senate Climate Bill Death. P. 7.

There will always be debates about the influence of money and politics. However, it is clear that there was an enormous amount of money provided by energy interests to Republican members of Congress in an attempt to influence the outcome of climate change legislation during 2009–2010. This does not consider prior contributions made by industry to members of Congress to influence the climate issue prior to 2009 and the hundreds of millions of dollars that was provided to coalitions for two decades that opposed any action to address climate change at the US and international levels. The amount of money cited above also does not include the amount the contributions provided by backers of the tea party.

An Overly Aggressive Agenda and Presidential Leadership

The book has provided data describing the precarious position of the US economy during 2008–2009. Because of this, President Obama and the Congress had a large to-do list. The $800 stimulus bill was adopted in February 2009. The debate over legislation to restructure the US health care system, the Patient Protection and Affordable Care Act, known as Obamacare, played out during 2009 and was signed into law on 23 March 2010. The Dodd-Frank Wall Street Reform and Consumer Protection Act, which overhauled the regulation of the US financial system was signed into law by President Obama on 21 July 2010. These pieces of legislation were all being debated at the same time. They were enormously complex, touching every aspect of the US economy and impacting all families.

The US Congress does not generally debate and pass such far-reaching legislation simultaneously. Many would have considered the 111th Congress productive based on the enactment of the stimulus package and Obamacare. At the time Obamacare passed in 2010, the US was spending $2.6 trillion, or nearly 18%, of GDP on health care.[115] As described, the cap-and-trade program and the hundreds of CPs included in the Waxman-Markey bill would have overhauled the US energy sys-

---

[115] The Henry J. Kaiser Family Foundation. *Health Care Costs: A Primer. Key Information on Health Care Costs And Their Impacts.* Number 7670–03, 2012. P. 1.

tem, which is enormous in scale and complexity. One analyst describes the passage of Obamacare and the effort to pass cap-and-trade as, 'a massive, complex, legislative effort to remake regulations, taxes, and expenditures affecting the entire US economy, all citizens, and powerful interest groups'.[116] This is a perfect characterization.

The quote, 'Never let a good crisis go to waste', was originally attributed to Churchill. Some believe that the Obama Administration adopted this philosophy in its attempt to advance its ambitious agenda. In short, my view is that there was too much on the agenda and it was moving too fast for the way in which Congress does business and for the public. Something had to give. It just so happens that climate change was it.

Some believe that climate change legislation could have passed if only President Obama exhibited greater leadership. I do not see any way in which the obstacles to the passage of legislation that have been cited could have been overcome. The one issue that is worthy of debate in this regard was if the Administration had put health care on the back burner and attempted to move climate change forward first. It is outside the scope of this effort to make such a determination.

The Liberals Write the Bill

Congressmen Waxman and Markey chaired the relevant committee's in the house with jurisdiction over energy and climate change. Because of the jurisdiction and long-standing association with environmental causes, it appears to make sense that they would be the two lead sponsors of climate change legislation. They did provide moderate members of the committee, such as Congressmen Boucher and Doyle, with opportunities to provide significant input into drafting key aspects of the bill.

In the bigger picture of climate change politics, Waxman and Markey's leadership roles may not have been the wisest course of action. They were viewed as two of the more liberal members of the house Democratic Caucus. The ambition of the legislation reflected this. Although they were both effective legislators with long records of accomplishments,

---

[116] Skocpol, T. Naming the Problem. P. 13.

their sponsorship was discomforting to much of industry and some Republicans. Chairman Waxman was also viewed by many as a strong partisan. One analysis cited his liberal record as a challenge. She wrote, '[t]he coalition wanted Waxman's long experience at crafting complex legislation, but it was wary of giving him so much ownership over the bill that it would scare away Republican members.'[117] The partisanship began right away when the Energy and Commerce Committee held a hearing on the USCAP's Blueprint the day of its release. And the role of Congressmen Waxman and Markey played may also have made it more difficult to secure the support of more moderate and conservative Democrats from the coal-reliant mid-west and southeast US and other energy-producing states. Similar to other factors that impacted the legislation's prospects for success, this issue may also have had an adverse impact.

The Inside Game Versus The Outside Game

The lobbying community that advocated for passage of the Waxman-Markey bill, including USCAP, focused its efforts predominantly on Washington, DC. It walked the floors of the house and senate office buildings, meeting with staff and members of Congress, making their case for the legislation, and providing detailed views on complex policy issues. This is known as 'traditional shoe leather lobbying'. Some call it an 'inside game'.

Some have argued that the failure to pass climate change legislation was due to advocates' emphasis on traditional Washington lobbying to pass the bill and their failure to create sufficient 'grass roots' support for it.[118] This approach is contrasted with advocates of health care legislation that created a coalition, Health Care for America Now, in States and districts to advocate for the bill and required its members to sign onto policy principles that should be incorporated in legislation. The grass

---

[117] Bartosiewicz, P., M. Miley, The Too Polite Revolution: Why the Recent Campaign To Pass Comprehensive Climate Legislation in the United States Failed. P. 39.

[118] This point was emphasized by both Skocpol and Bartosiwiecz and Wiley in their work.

roots approach is cited by some as having been critical to the passage of health care legislation.[119]

There are different types of grassroots. There is traditional 'astro turf' grassroots which has been used for decades. It is generally defined as constituents providing their views to elected officials through conventional means such as postcards, letters, and phone calls to their district or Washington offices or in rallies or town hall meetings. Many of these campaigns are orchestrated by firms that are hired for this purpose. Although this may have changed, based on my previous experience as a lobbyist, I believe these campaigns' effectiveness is partially based on whether the elected officials believe constituents are communicating genuine concerns or are the result of 'hired guns' efforts.

Another type of grass roots is called 'grass tops'. I am much more familiar with 'grass tops' as my company employed this approach on behalf of a client before it was widely used. Grass tops is a process in which prominent individuals in a State or district, frequently business, labor, environmental, or civics leaders with close ties to an elected official, are enlisted to communicate their views on an important public policy issue. The communication is typically in the form of a brief note, phone call, or personal communication. The idea behind 'grass tops' was that members of Congress pay close attention to the views of local opinion leaders, many of whom they have long-standing relationships with, in their consideration of an issue. We employed this strategy on bread-and-butter issues of taxation. It is easy to communicate the economic impacts of taxes in a 30-second conversation or a note. In contrast, issues like climate do not appear to lend themselves to such an approach.

It is difficult to assess whether climate change legislation may have passed if a comprehensive grass roots program was put in place to support it. I am doubtful that such an effort could have moved the needle at all. This is because creating grassroots support for an issue as complex as climate change is exceptionally difficult. It has not been done effectively since concerns with the issue began nearly a quarter of a century ago. Advocacy groups, such as Al Gore's Alliance for Climate Protection and Clean Energy

---

[119] See pages 34–54 in Skocpol.

Works, did invest in advertising in support of climate change legislation during 2009–2010. It does not appear to have aided the cause.

Why is it so challenging to create effective grassroots support for climate change legislation? Doing so requires the mobilization of citizens to communicate policy benefits to policy-makers. They need to be convinced to take time out of their busy schedules to do so. If this is possible, success is difficult for several reasons. The first is that simple concise messages are required to effectively communicate policies benefits to policy-makers. Trained professionals have found it difficult to communicate the benefits of legislation such as mitigating the risk of climate change for future generations, improving energy security, or creating green jobs. It would be as difficult, or more so, for citizens. And it would be extremely hard to do in the best of times. In the middle of a significant economic downturn such as when the climate change debate was taking place, I do not know if it is possible.

This is in contrast to creating grassroots opposition to climate change legislation. The opponent's messages regarding the impacts of cap-and-trade legislation or a national energy tax, as it has been frequently characterized, including job loss, increased electricity rates and gasoline prices, and the creation of a complex carbon market that would enrich Wall Street bankers, resonate. These impacts are personal, making it easier to mobilize workers in the traditional energy and manufacturing industries and citizens located in potentially affected communities to communicate with policy-makers. These messages provide the energy to develop an effective grassroots campaign to oppose climate change legislation. And that is exactly what opponents did. Americans for Clean Coal Electricity sent workers to 264 state fairs, Kiwanis meetings, and college campuses; the American Petroleum Institute funded rallies; and Americans for Prosperity, founded and funded by the billionaire Koch brothers, held dozens of events against cap-and-trade.[120] These groups also used advertising to communicate their messages. These techniques have worked well for opponents of climate change policy for 25 years. It has always been easier to stop things in Washington than to advance them. This is

---

[120] Layzer, J. (2011) 'Cold Front' in Skocpol, T., L. Jacobs (eds.) *Reaching for a New Deal.* (New York, NY. Russell Sage Foundation). PP. 321–385.

magnified by the differences in communicating the simple and personal messages that opponents of legislation have at their disposal in contrast to supporters attempting to communicate the benefits of action.

The characteristics of the health care issue are similar to those that enabled the creation of effective grassroots campaigns against climate change legislation. Millions of Americans have been affected by increased health care costs and fear of not being able to take care of a family member. Similarly, opponents of health care legislation can be motivated by concerns that legislation would increase their insurance costs or change their relationships with their physicians. These concerns provide the ability to motivate citizens to participate in grassroots efforts for or against health care reform and to communicate personal and hard-hitting messages to policy-makers. This is in contrast to climate change legislation. It is easy to motivate opponents. Supporters are left attempting to communicate that legislation will slow climate change, translate to more green jobs, and improve national security—benefits that seem illusory to some.

A Top-Down, Overly Ambitious, Complex Piece of Legislation

All of the dynamics cited above played a role in ACES' defeat and the end of climate change 1.0 in the US. The legislation officially died in the senate. Its prospects ended on 26 June 2009, the day ACES passed the US House of Representatives. The legislation's supporters over-reached. I don't know if this was caused by misreading the public's support for legislation, if they saw an opening due to the ongoing economic crisis, or believed they could slam the bill through the Congress will little or no Republican support because of the large Democratic majorities in the house and senate.

Most of the analyses and postmortems regarding the bill's demise have one thing in common: they fail to focus on the substance of the legislation. And supporters' failure to focus on these issues will result in similar mistakes in the future. The bill passed by the house was so top-down, bureaucratic, overly ambitious, and complex that it was never realistic to expect it or anything similar to it to become law. In my view, these characteristics of the legislation were as responsible for its defeat or more so than any of the reasons that have been cited.

American Clean Energy and Security Act of 2009's Ambition and Complexity

We have tracked ACES' growth as it advanced in the legislative process. It grew from 648 pages in March 2009,[121] to the 932-page bill marked up by the Energy and Commerce Committee on 21–24 May 2009,[122] and to 1428 pages when passed by the house.[123] It grew by 780 pages in less than 90 days!

As described earlier in this chapter, ACES included a cap-and-trade system covering nearly 85% of US GHG emissions, included approximately 25 different allowance allocations valued in the hundreds of billions of dollars that were used to fund multiple objectives, an array of CPs designed to remake the US energy system, and numerous other provisions. There is no guarantee the carbon market would have worked as intended given the inclusion of so many CPs in the legislation. Its performance could have been similar to the EU ETS. The interaction of the CPs and the market were not discussed thoroughly during the development of the legislation.

Implementation would have been resource intensive and complex. I previously quoted one member of Congress' statement on the house floor citing a Chamber of Commerce study that 397 traditional rulemakings were required to implement the programs included in the legislation.[124] Development and implementation of nearly 400 rules would take years—if not decades—and require significant financial and human resources.

The legislation played into the Republican narrative regarding Democrats being the party of tax-and-spend, big government, and bureaucracy. It provided ACES opponents with simple messages that

---

[121] Available from: https://wayback.archive-it.org

[122] Available from: https://wayback.archive-it.org

[123] Available from: https://wayback.archive-it.org

[124] See statement made by Representative Dan Burton in the Congressional Record, P. H7666, 26 June 2009.

were easy to communicate and which they used effectively to mock and attack the entire effort. Although the success of the Republican attacks on Democrats waxed and waned over time, they were more likely to be effective during bad economic times.

What follows are a description of a few provisions in ACES which illustrate its ambition, complexity, and heavy handedness that helped to create the narrative, contributing to its defeat and the end of climate change policy 1.0 in the US.

The Global Warming Title

1. Coverage

The cap-and-trade program would have covered nearly 85 % of national GHG emissions. Such programs under the CAA had typically covered stationary sources such as power plants and large manufacturing facilities that were responsible for approximately 50 % of US emissions. Many of the owners and operators of these facilities were familiar with cap-and-trade through their participation in the US acid rain program and others authorized by the CAAA. ACES would have expanded coverage of the cap-and-trade system to emissions created by the combustion of fossil fuels in the transportation sector. In theory, expanding the program to cover a larger share of GHG emissions was a net plus for the market. The argument being that covering a higher percentage of emissions would have led to a larger, more efficient market and greater opportunities for cost-saving trade.

At a meeting on the potential to cover transportation sector emissions in a cap-and-trade program several years prior to 2009, a friend of mine who was an executive at a utility company and had been a staff member of the House Energy and Commerce Committee and the EPA leaned over to me and said this would not work. He said cap-and-trade should cover stationary sources and transportation sector emissions should be addressed by the corporate average fuel economy (CAFÉ) program that had been created decades earlier.

His reasoning was that this approach would have been more politically acceptable to policy-makers, the regulated community, and other participants because of their familiarity with the program. In short, he was right. The effort to cover the transportation sector emissions increased the program's ambition and complexity and the sectors' political opposition to the legislation.

2. Allocations

An enormous amount of energy was spent attempting to address important substantive concerns and to create political support for the legislation through the allocation of allowances. These allocations, which were described earlier, shielded energy consumers, trade-sensitive industries, and low- and middle-income households from increased energy costs, and financed the development and deployment of low- and non-emitting energy technologies.[125] Allowances were also allocated to fund many other objectives including tax cuts and to avoid deforestation projects outside of the US.[126] These allocations subjected the legislation to ridicule.

To illustrate the legislation's ambition, ACES included approximately 25 allowance allocations, many of which were used to pay for favored programs. They were used by supporters, similarly to the ways taxes or spending were proposed to be used, to fund projects or programs important to individual members of Congress. Consider this quote from Senator Boxer, who was chairing the Senate Committee on Environment and Public Works in 2009, in referring to the house process. 'I believed he could do it', Boxer said of Waxman (chairman of the Energy and Commerce Committee and the legislations chief sponsor). 'There's so much revenue that comes in from a cap-and-trade system that you can really go to a person in a congressional district and get enough votes there by saying, "What do you need? What do you want? You can really

---

[125] For a review of these allocations, see Proposed Allowance Allocation. Available from: https://wayback.archive-it.org

[126] All allocations can be found in Appendix A, Allocation of Emission Allowances in a summary of the house passed bill by the Pew Center on Global Climate Change, now the Center for Climate and Energy Solutions. Available from: http://www.c2es.org/docUploads/Waxman-Markey%20summary_FINAL_7.31.pdf [Accessed 18 December 2015].

help them." '[127] Republicans' views on this were pretty well summed up by Senator Bob Corker (R-TN) who said, 'God, it's just a public bribery bill, isn't it?'[128]

In the end, the allocations caused several problems. Those made to protect energy consumers, low-income groups, and manufacturers from rising energy costs validated opponents' views that cap-and-trade was a tax that would raise energy prices and cause a loss of jobs. If it would not, why was there a need for the allocations to guard against this outcome? This enabled the legislation to be labeled as 'cap and tax' or as a 'national energy tax' that would increase energy prices and destroy jobs.

Allocations and other mechanisms included in the bill to stimulate investment in energy technologies alone were valued at nearly $200 billion,[129] providing the Republican with ammunition to characterize the legislation as a large new government program and its supporters as 'tax and spend'. Analysis of ACES undertaken by CBO and the Joint Committee on Taxation, which is comprised of members of Congress and Senators from each body's tax-writing committees, concluded that the legislation would raise $873 billion from 2010 to 2019 and increase spending by $864 billion during the same period.[130] So, although the legislation reduced the deficit by approximately $9 billion, it raised and spent an enormous amount of revenue. It provided the Republicans with further ammunition to portray the legislation's sponsors as 'tax and spend'.

The charge that cap-and-trade was equivalent to an energy tax that would destroy jobs and adversely impact the economy had the greatest impact on its prospects. The similarities in the opponents' messages used to defeat ACES and President Clinton's BTU tax are remarkable. Identical to Republican characterizations of ACES on the house floor, the industry coalition established to oppose the BTU tax called it a national

---

[127] Simendinger, A. (2009) 'Will key chairman power up for energy bill?' *National Journal.*

[128] Simendinger, A. (2009) 'Will key chairman power up for energy bill?' *National Journal.*

[129] This amount is included in a July 2009 summary of the bill passed by the House of Representatives by the House Committee on Energy and Commerce. Available from: https://wayback.archive-it.org

[130] Congressional Budget Office and the Joint Committee on Taxation. *Estimated Changes In Revenues And Direct Spending Under HR 2998, As Amended and Reported by the House Committee on Rules On June 26, 2009.* 26 June 2009.

energy tax or a job-destroying energy tax.[131] This message is simple, hard-hitting, and effective. Hit people in their pocketbooks. And in another unfortunate similarity between ACES and the BTU tax, supporters failed in their attempt to effectively communicate the benefits of their proposals. As communicated in one postmortem of the failure to pass ACES, '[t]he concept of green jobs was difficult to rally behind largely because creating green jobs was as abstract as reversing climate change'.[132]

The politics was also the same. The Republicans saw an enormous political opportunity to defeat supporters of ACES in the upcoming 2010 election. When the votes were read, Republicans on the floor even shouted 'B-T-U, B-T-U'.[133] Republicans remembered that the BTU tax was considered a factor in the defeat of 27 house Democrats in the 1994 elections.[134] The Democrats lost control of the house in 1994. And based on my experience in that issue, the BTU tax certainly contributed to the Democrats' defeat. Republicans also took back control of the house following the 2010 elections. Whereas many issues contributed to this outcome, cap-and-trade did not help the Democrats' cause.

The allocations also helped to create the argument that companies receiving them were feeding at the trough and recipients of corporate welfare. Once again, the CBO findings that 50% of allowance value went to business played into the Republican narrative of the legislation.[135]

All of these attacks were included in former Speaker Gingrich's testimony and others cited previously. In short, the allowance allocation process over-reached in its attempts to solve problems and buy political support. It resulted in providing the legislation opponents with simple arguments that played on the publics' fears during difficult economic times and were never effectively refuted.

---

[131] Erlander, D. (1994) 'The BTU Tax Experience: 'What Happened and Why it Happened', *Pace Environmental Law Review,* 12(1): 179.

[132] Bartosiewicz, P., M. Miley. The Too Polite Revolution: Why the Recent Campaign To Pass Comprehensive Climate Legislation in the United States Failed. P. 59.

[133] Bartosiewicz, P., M. Miley. The Too Polite Revolution: Why the Recent Campaign To Pass Comprehensive Climate Legislation in the United States Failed. P. 49.

[134] See footnote 11 on P. 12 in Bartosiewicz and Wiley.

[135] Congressional Budget Office. *The Estimated Costs to Households.*

3. Strategic Reserve

Out of concern with high allowance prices, a strategic reserve was created. It authorized covered entities to purchase allowances from the reserve at minimum prices in auctions, capping compliance costs. The legislation required the proceeds from strategic reserve allowances sales to be used to purchase international offsets created by avoided deforestation, or as known in the bill, supplemental emission reductions through reduced deforestation. The offsets would be retired and used to establish allowances equal to 80 % of the retired offsets.[136] Although well intentioned, I always found this provision to be confusing.[137]

4. Auction

The flip side of high prices is low ones—ones that do not send adequate price signals for investment. Quarterly auctions were established and required that allowances be purchased at a minimum price.[138] And as an added benefit, the proceeds from auctions sales were used to fund several programs including tax cuts.

In my view, these provisions also increased the legislation's complexity. This is because they raised money from allowance sales and used it to finance other priorities.

Carbon and Energy Market Regulation

There was concern that cap-and-trade legislation would create a large new commodity market, susceptible to speculation and market manipulation. The provisions in ACES creating a regulatory program to guard against these outcomes were highlighted earlier in the chapter. Few understood them when they were being created. Much of the substance of market regulation was subject to the jurisdiction of other house committees.

---

[136] The Strategic Reserve can be found in Title VII, Part C. Available from: https://wayback.archive-it.org

[137] The provision creating the supplemental emission reductions through reduced deforestation can be found in Title VII, Part E. Available from: https://wayback.archive-it.org

[138] The provision creating the auction can be found in Title VII, Part H. Available from: https://wayback.archive-it.org

The potential scale and complexity of the carbon market that the legislation would create raised concerns and provided Republicans with a new line of attack. They compared it with the housing market, which many believed was a major cause of the economic downturn in the US. They argued that bankers, Wall Street, and other bad guys would create exotic financial instruments and would enrich themselves at the expense of the public. Republicans sounded like Democrats! Could this have been avoided? May be not. But it could have been easier to refute if it was not as complicated.

The point in highlighting some of the provisions in ACES' global warming title, and others to address related concerns, is to illustrate the legislation's ambition and complexity and the lines of attack it provided to ACES opponents. This was partially a result of the legislation's sponsors attempting to develop a solution for every potential problem that could be created by a cap-and-trade program and resultant carbon market and to fund favored programs. We see this throughout ACES in provisions that use allowance allocations to shield various groups against energy price increases and to fund technology, and to use the proceeds created by mechanisms to guard against high and low allowance prices to fund tax cuts and to prevent international deforestation. And because the legislation created a large new market, which opponents gleefully compared to the housing market, a new system of regulation had to be developed and incorporated in the bill. Unfortunately, this all contributed to the legislation's defeat.

Clean Energy and Energy Efficiency

Some of the CPs included Titles I and II of the legislation were briefly described previously. These titles had grown to almost 700 pages! What follows is but one example of a program that contributed to the legislation's defeat.

Title II authorized the DOE to promulgate efficiency standards for lighting and many other products. Legislation providing the federal government with the authority to set performance standards for energy-using products has been a cornerstone of US energy policy since the mid-1970s. These programs have been successful in reducing energy consumption, spurring new technologies, and reducing emissions of CO_2 and air pollutants. Standards are an effective form of regulating these products, given that they are widely disbursed in millions of residential, commercial, and industrial buildings throughout the US. President Obama has used these programs effectively to achieve significant reductions in GHG emissions and to achieve other objectives.

The standards programs were ridiculed by ACES' opponents. It should have been anticipated that mandates imposing standards on a multitude of products and the other programs in Titles I and II that filled hundreds of pages combined with cap-and-trade would subject it to significant attack. The legislation's opponents would use these provisions that were predominantly regulatory in nature and others to argue that ACES was top-down big government-run amok that would intrude into citizen's liberties, result in a takeover of energy policy by the federal government, grow regulation, and increase the power of federal bureaucrats at the expense of States and the public. Same old Democrats! And the specific references requiring that standards be imposed on products such as water dispensers, hot-food-holding cabinets, and portable electric spas provided Republican opponents with a sound bite to ridicule the entire effort. This is exactly what former Speaker of the House Gingrich did in his testimony of the legislation when he said the Secretary of Energy was going to be the 'Czar of the Jacuzzis'. Many of his Republican colleagues reinforced these attacks on the house floor with the reference to bureaucratic nature of the legislation and the 397 rulemakings required to implement it. And although I only referenced the standards programs, others included in the clean energy and energy efficiency titles of ACES could have been subjected to similar attacks.

Any legislation that would create nearly 1500 pages of new law is going to be complex. There is no way to avoid this. But it had an adverse effect. Michael Paar, a senior manager of DuPont, and a key member of USCAP said, 'Hell, I barely understood the bill and I basically wrote it.'[139] In

---

[139] Bartosiewicz, P., M. Miley, The Too Polite Revolution: Why the Recent Campaign To Pass Comprehensive Climate Legislation in the United States Failed. P. 37.

another quote, he said the complexity of the bill was driving away its staunchest proponents.[140] 'We could no longer really look at the bill and understand what its economic impact would be on our operations.'[141] And this was a supporter from a leading USCAP company. The legislation's top-down heavy-handed approach, ambition, and complexity provided opponents with simple messages to attack the bill and provided a road map to kill it.

Similarities in the Causes of Failure

There were many differences in the causes of failure of climate change 1.0 in the US and at the international level. However, there were also similarities that should not be ignored in ongoing and future efforts to reduce GHG emissions. The KP and ACES were top-down, bureaucratic, overly ambitious, and complex. The attributes common to climate change policy 1.0 in the US and the international level led to their demise and ushered in a new era of policy-making. In baseball parlance, the authors of ACES and the KP attempted to hit a home run. They might have been more successful with singles and doubles. For example, we will never know whether it would have been possible to pass a cap-and-trade bill limited to the power sector in the US.

These are the lessons I have drawn from a quarter of a century of climate change policy-making. The most important issue in this regard is how they are applied to future efforts. Experience gained from the first generation of policy-making has changed my views as to the philosophy that should guide the development of future policy-making efforts and the policies that would constitute an effective, enduring response to climate change. I will attempt to articulate this philosophy and recommend policies for climate change 2.0 in Chap. 7.

---

[140] Bartosiewicz, P., M. Miley, The Too Polite Revolution: Why the Recent Campaign To Pass Comprehensive Climate Legislation in the United States Failed. P. 62.

[141] Bartosiewicz, P., M. Miley, The Too Polite Revolution: Why the Recent Campaign To Pass Comprehensive Climate Legislation in the United States Failed. P. 62.

6

US and International Climate Change 2.0 Emerges

If society is to successfully address climate change, a new approach is required. It is time to move away from the ambitious attempts that were the corner stone of climate change 1.0. More modest, flexible, and workable polices need to be put in place in climate change 2.0 in both the US and internationally.

This chapter reviews the status of climate change 2.0, with an emphasis on developments in the US, to achieve reductions of GHG emissions and the international community's efforts to develop a successor to the KP.

US Climate Change Policy 2.0

Given the leadership role the US plays in the world, and that it remains the largest economy and second largest emitter, its failure to enact a coherent climate change strategy during the last 25 years has been disappointing, to say the least. Failure culminated in the defeat of ACES. However, signs of climate change 2.0 are seen as early as 2009 in the US, when the Obama Administration began to use existing laws to reduce GHG emissions. This

© The Author(s) 2016   **211**
R.H. Rosenzweig, *Global Climate Change Policy and
Carbon Markets*, Energy, Climate and the Environment,
DOI 10.1057/978-1-137-56051-3_6

effort grew into a strategy whose cornerstone relied on far-reaching regulations using the decades-old CAA to achieve emission reductions from the transportation and power sectors, the two largest emitting sectors in the US. The success of the strategy ultimately rests on the outcome of the legal challenges and political attacks of its most visible component, the Clean Power Plan (CPP). In addition to describing some of the key elements of President Obama's Climate Action Plan (CAP), this section highlights some of the prominent subnational efforts in the US that have been put into place. What follows is a brief description of the approach the US is taking in climate change 2.0 and a discussion of whether it avoids the mistakes of the past and can be scaled up.

US GHG Emissions Targets

In 2009, in a speech delivered at COP-15 in Copenhagen, President Obama committed the US to reduce its GHG emissions to 17% below 2005 levels by 2020.[1] This was consistent with the target included in ACES, which had passed the US House of Representatives approximately six months earlier. Five years later, and in the context of announcing an agreement between the US and China on climate change and clean energy, President Obama committed the US to reducing its net GHG emissions to 26–28% below 2005 levels by 2025.[2] Achieving the more ambitious target would require doubling the pace of reductions to approximately 2.3–2.8% per year from 2020 to 2025 from the reductions required to achieve the prior target. And to illustrate how much has changed in the US, the primary advocate of carbon markets in climate change 1.0 communicated in its intended nationally determined contribution (INDCs) that it would not use international market mechanisms to achieve its target.[3]

---

[1] Obama, President B. Remarks by the President at the Morning Plenary Session of the United Nations Climate Change Conference. Copenhagen, Denmark. 18 December 2009.

[2] The White House Office of the Press Secretary. U.S.–China Joint Presidential Statement on Climate Change. Available from: https://www.whitehouse.gov/the-press-office/2015/09/25/us-china-joint-presidential-statement-climate-change [Accessed 21 December 2015].

[3] United States of America. US Intended Nationally Determined Contribution. Available from: http://www4.unfccc.int/submissions/INDC/Submission%20Pages/submissions.aspx [Accessed 21 December 2015].

Federal Policies

Although the president's CAP was not communicated until June 2013,[4] the Administration had taken many climate-related actions beginning in 2009.[5] The Administration moved quickly to undertake the analysis required by the Supreme Court in Massachusetts vs. EPA. The analysis, which became known as the endangerment finding, assessed the extent to which GHG emissions threatened public health and welfare.[6] In December 2009, the EPA administrator signed the endangerment finding, which concluded six GHGs 'in the atmosphere threaten the public health and welfare of current and future generations'.[7] The administrator also found 'that the combined emissions of these well-mixed greenhouse gases from new motor vehicles and new motor vehicle engines contribute to the greenhouse gas pollution which threatens public health and welfare'.[8] These findings laid the groundwork for President Obama's ambitious regulatory agenda.

What follows are highlights of some of the key policies that the Administration put in place to achieve its GHG emissions reduction target in 2025.

Transportation

Achieving GHG reductions from the transportation sector is critical given that it is the second largest emitting sector in the US in 2013, responsible for 27% of national emissions.[9] The largest contributor to these emis-

---

[4] Executive Office of the President. *The President's Climate Action Plan.* The White House. 2013.

[5] For a list of actions taken by the administration see Table 8A.1. in Layzer J. (2011) 'Cold Front: How the Recession Stalled Obama's Clean Energy Agenda' in Skocpol, T., L. Jacobs (eds.) *Reaching for a New Deal: Ambitious Governance, Economic Meltdown, and Polarized Politics in Obama's First Two Years* (New York, NY. Russell Sage Foundation). PP. 321–385.

[6] US Environmental Protection Agency. *Endangerment and Cause or Contribute Findings For Greenhouse Gases under Section 202 (a) of the Clean Air Act.* Federal Register. 74(239): 2009.

[7] US EPA. *Endangerment Finding.*

[8] US EPA. *Endangerment Finding.*

[9] US EPA. Sources of Greenhouse Gas Emissions. Available from: http://www3.epa.gov/climatechange/ghgemissions/sources.html [Accessed 21 December 2015].

sions is light-duty vehicles, which account for 60 % of the total, followed by medium- and heavy-duty vehicles, which account for nearly a quarter of the sector's emissions.[10]

Light-Duty Vehicles

The Administration completed two rules imposing GHG standards on light-duty vehicles using the Clean Air Act (CAA) and reducing oil consumption under the Corporate Average Fuel Economy (CAFÉ) program included in the Energy Policy and Conservation Act. The first rule, completed in 2010, applies to model years 2012–2016 and imposed a standard for CO_2 tailpipe emissions and fuel economy.[11] The second rule, completed in 2012, applied to model years 2017–2025 tightened the standard for both CO_2 tailpipe emissions and fuel economy.[12] The Administration estimates that these rules would cut GHG emissions significantly over the life of model years 2012–2025, while reducing fuel costs and oil consumption dramatically.[13]

Medium-Duty and Heavy-Duty Vehicles

In 2011, the Administration completed another rule using its authority under the CAA and the Energy Security and Independence Act to impose GHG emission standards and fuel consumption standards on medium- and heavy-duty vehicles such as commercial trucks, vans, and buses for model years 2014–2018 and engines.[14] The Administration estimates

---

[10] US EPA. *Fast Facts: US Transportation Sector GHG Emissions 1990–2011.* Office of Transportation and Air Quality. EPA-420-F-15-032. 2015.

[11] US EPA, US Department of Transportation. *Light-Duty Vehicle Greenhouse Gas Emissions Standards and Corporate Average Fuel Economy Standards; Final Rule.* Washington. 40 CFR Parts 85, 86, and 600. 2010.

[12] US EPA, US Department of Transportation. *2017 and Later Model Year Light Duty Vehicle Greenhouse Gas Emissions Standards and Corporate Average Fuel Economy Standards; Final Rule.* Washington. 40 CFR Parts 85, 86, and 600. 2012

[13] US EPA. *Regulation and Standards: Light Duty.* Available from: http://www3.epa.gov/otaq/climate/regs-light-duty.htm [Accessed 21 December 2015].

[14] US EPA, US DOT. *Greenhouse Gas Efficiency Standards and Fuel Efficiency Standards for Medium- and Heavy-Duty Engines and Vehicles; Final Rule.* Washington. 40 CFR Parts 85, 86, 600, 1033,

that the rule would cut GHG emissions and fuel use and result in net benefits of approximately US$50 billion.[15] It has proposed a Phase 2 rule for this category of vehicles applicable to post-2020 model years requiring further improvements to fuel efficiency and GHG emissions.[16]

The EPA also made a preliminary endangerment finding regarding GHG emissions from the aviation sector which account for 8 % of transportation sector GHG emissions.[17, 18] Similar to the endangerment findings that opened the door to the rules cited above, an identical finding was required to regulate the aviation sectors' emissions. If EPA moves forward, approximately 90 % of transportation sector emissions would be subject to regulation.

The Power Sector

The power sector is the largest emitting sector in the US, accounting for 31 % of national GHG emissions.[19]

The Clean Power Plan (CPP)

The most notable policy put forward to date by President Obama to reduce GHG emissions is the Carbon Pollution Emission Guidelines for Existing Stationary Sources: Electric Utility Generating Units which

---

1036, 1037, 1039, 1065, 1066, and 1068. 2011.

[15] US EPA. *EPA and NHTSA Adopt First-Ever Program to Reduce Greenhouse Gas Emissions and Improve Fuel Efficiency of Medium- and Heavy Duty Vehicles.* Available from: http://www3.epa.gov/otaq/climate/documents/420f11031.pdf [Accessed 21 December 2015].

[16] US EPA, US DOT. *Greenhouse Gas Emissions and Fuel Efficiency Standards for Medium and Heavy-Duty Engines and Vehicles—Phase 2; Proposed Rule.* 40 CFR Parts 9, 22, 85, 86, 600, 1033, 1036, 1037, 1039, 1042, 1043, 1065, 1066, and 1068. Washington. 2015.

[17] US EPA. Proposed Finding That Greenhouse Gas Emissions From Aircraft Cause or Contribute to Air Pollution That May Reasonably Be Anticipated To Endanger Public Health and Welfare and Advance Notice of Proposed Rulemaking. Washington. 40 CFR Parts 87 and 1068. 2015.

[18] US EPA. *Fast Facts: US Transportation Sector GHG Emissions.*

[19] US EPA. *Sources of Greenhouse Gas Emissions. Estimates from the Inventory of US Greenhouse Gas Emissions and Sinks: 1990–2013.* Available from: http://www3.epa.gov/climatechange/ghgemissions/sources/electricity.html [Accessed 21 December 2015].

is known as the CPP.[20] When fully implemented in 2030, the CPP is designed to achieve a 32% reduction in power sector emissions below 2005 levels. This policy constitutes the exception to my view regarding the need to avoid ambitious, politically contentious policies. Given that taxes and cap-and-trade are off the table in the US, regulation is the only way to achieve large-scale reductions of GHG emissions from this sector. And there is no way to tackle climate change in the US without achieving reductions from electricity generation.

The regulation requires reductions in three phases over an eight-year period from 2022 to 2024, 2025 to 2027, and 2028 to 2029. Each state is assigned an emission reduction target. As is customary under the CAA, states are required to submit a plan to EPA detailing how they will meet the target. If they do not, the EPA put forward a model federal plan that would require each unit in a state to meet a target.

The rule is designed to provide each state with the flexibility to determine how to achieve its GHG emission target. It is expected that the state compliance plans would include improving the supply-side efficiency of power plants, fuel switching from coal to natural gas, increasing the use of renewables, and demand-side efficiency. Some states and regulated firms have expressed an interest in utilizing emissions trading as a means to comply with the GHG reduction targets. This is particularly ironic given the defeat of legislation that would have established a larger, more liquid market and had the legitimacy of congressional action. The rule provides a path for states and regions that would like to use trading as a compliance strategy.

EPA has widely publicized the climate, air quality, public health, and energy benefits that would result from implementing the rule. The rule's 32% reduction requirement translates to 870 million tons less of carbon annually by 2030, which is equal to the annual emissions of 166 million cars and the emissions from annual residential electricity use in the US. The rule would also help create significant public health benefits resulting from reductions in emissions of SO_2 and NOx, which cause soot and smog. Net benefits of the rule are assumed to be between US$26–

[20] US EPA. *Carbon Pollution Emission Guidelines for Existing Stationary Sources: Electric Utility Generating Units*, Final Rule. Washington. 40 CFR Part 60. 2015

US$45 billion in 2030. And, renewable energy is estimated to grow from 12% of the US energy use in 2012, to 21% in 2030, a 75% increase.[21]

In addition to the requirements imposed on existing power plants, EPA also completed a rule that sets an emission standard for new, modified, and reconstructed fossil fueled power plants.[22]

The CPP is a historic rule. It set off a political firestorm. Litigation was filed the day the final rule was proposed. Senator Mitch McConnell (R-KY), the Majority Leader of the US Senate, wrote a letter to the National Governors Association on 19 March 2015 urging the governors not to file the required compliance plans with EPA.[23] In more than 30 years of involvement in federal energy and environmental policy, I have never heard of a congressional leader urging such a course of action. Many governors will disregard this advice, although some will not.

Given the unpredictable twists and turns of litigation, the rule's future is, and will remain, highly uncertain for several years. If it does take effect, its final form may be different because of changes required by the courts. The rule will also be subject to continuous political attacks from members of congress, primarily Republicans, who will attempt to stop its implementation. In addition, if a Republican wins the upcoming presidential election, he or she could attempt to dismantle the rule, although this would require the use of a significant amount of political capital. Or, a Republican president could allow its implementation and blame any adverse impacts on President Obama. The legal and political attacks on the rule indicate why legislation would have been a better solution. As stated earlier, regulations do not have the same legitimacy in a democracy as legislation.

Following the defeat of ACES, President Obama did not have any other options to address climate change but to move forward with such

---

[21] Impacts of the Clean Power Plan can be reviewed in several fact sheets on the rule developed by the US EPA. Available from: http://www.epa.gov/cleanpowerplan/clean-power-plan-existing-power-plants [Accessed 21 December 2015].

[22] US EPA. *Standards of Performance for Greenhouse Gas Emissions from New, Modified, and Reconstructed Stationary Sources: Electric Utility Generating Units; Final Rule.* Washington. 40 CFR Parts 60, 70, 71, and 98. 2015.

[23] Senator McConnell's letter to the National Governors is available from: http://www.mcconnell. senate.gov/public/index.cfm?p=newsletters&ContentRecord_id=d57eba06-0718-4a22-8f59-1e610793a2a3&ContentType_id=9b9b3f28-5479-468a-a86b-10c747f4ead7&Group_id=2085dee5-c311-4812-8bea-2dad42782cd4 [Accessed 21 December 2015].

an approach. It is not possible to reduce national GHG emissions without tackling GHG emissions from the power and transportation sectors. Combined, they account for nearly 60 % of US emissions.

Other Policies

What follows is a brief description of other policies being utilized by President Obama to reduce US GHG emissions.

Renewable Energy

Transitioning to a clean energy economy was one of President Obama's most ambitious campaign commitments. The 2013 CAP established a goal of doubling renewable generation by 2020.[24]

The administration has put several programs in place to increase the use of renewables beginning with ARRA. An array of regulations, research and development spending, policies designed to pull them into the market such as tax incentives, and to permit projects on federal lands are encouraging increased use of renewables.

It is difficult to determine the impacts of individual policies in increasing renewables use. However, by any measure, renewables have made impressive gains since the beginning of President Obama's presidency. Non-hydro renewable use in the US increased from 41 gigawatts in 2008 to 103 gigawatts in 2014[25]—an increase of 2.5 times, while increasing its share of US electricity supplies from 8 to 13 %.[26] Wind power and solar energy made significant gains during this time period.[27] These trends are forecast to continue. The EIA estimates renewables will add nearly 40 % of all new generation in the US until 2040.[28]

---

[24] Executive Office of the President. *The President's Climate Action Plan.* P. 6.

[25] Zindler, E., M. Di Capua et al. *2015 Factbook: Sustainable Energy in America.* Bloomberg New Energy Finance, The Business Council for Sustainable Energy. 2015. P. 19.

[26] Zindler, E., M. Di Capua et al. *2015 Factbook: Sustainable Energy in America.* P. 7.

[27] Zindler, E., M. Di Capua et al. *2015 Factbook: Sustainable Energy in America.* P. 19.

[28] US Energy Information Administration, *Annual Energy Outlook 2015 with projections till 2040.* US DOE. DOE/EIA – 0383(2015), 2015. P. 26.

Energy Efficiency

Another key energy goal of the Administration, and important to reducing GHG emissions, is to improve the energy efficiency of the US economy. The president's CAP included a national goal to reduce carbon emissions by a cumulative three billion metric tons by 2030 through new efficiency standards for appliances and federal buildings.[29]

The Administration has been aggressive on the efficiency front, particularly in establishing standards for energy using equipment such as lighting and appliances and will continue to be. The DOE has used existing authority to issue more than 30 new or updated standards for many different types of equipment, and the Administration has committed to finalize more than 20 additional ones by the end of 2016.[30,31, 32] In addition, other policies including ARRA and the Clean Power rule will also improve efficiency in the power sector.

Importantly, the efforts to increase investment in renewables and to improve energy efficiency to reduce GHG emissions will provide other economic, energy, security, and environmental benefits.

Measures to Reduce Non-CO$_2$ Gases

The rules targeting emissions from the transportation and power sectors are the largest programs that have been developed during the Obama Administration to reduce national GHG emissions. The renewables and efficiency programs will also make a contribution to the US achieving its emission reduction goals and improvements in carbon and energy intensity. Although less publicized,

---

[29] Executive Office of the President. *The President's Climate Action Plan.* P. 9.

[30] US Department of State. *United States Climate Action Report 2016, Second Biennial Report of the United States of America, Under the United Nations Framework Convention on Climate Change.* 2016. P. 19.

[31] The White House Office of the Press Secretary. *US–China Joint Presidential Statement on Climate Change.*

[32] The administration has communicated that the efficiency standards finalized by 2015 are estimated to avoid more than 2.2 billion MT of carbon emissions by 2030. See Executive Office of the President. *President Obama's Climate Action Plan, 2nd Anniversary Progress Report,* The White House. 2015. P. 8.

the Administration has also used its existing authority to develop rules to control emissions of non-CO_2 gases including methane, HFCs, and others that comprise 18% of US emissions.[33]

Methane

The Administration committed to develop a strategy to address methane, which comprises 10% of US emissions.[34, 35] It proposed a series of rules designed to achieve reductions from the oil and gas sectors and land-fills.[36, 37]

Hydrofluorocarbons (HFCs)

The Administration also committed to reducing emissions of HFCs,[38] which comprise a small but growing percentage of US GHG emissions.[39, 40] It used existing authority under the Significant New Alternatives Policy Program (SNAP), which allows and disallows the use of certain chemicals.[41] Internationally, the US proposed amendments to the Montreal Protocol in 2015 to phase out HFC use.[42]

---

[33] US EPA. *U.S. Greenhouse Gas Emissions in 2013*. Available from: http://www3.epa.gov/climatechange/ghgemissions/gases.html [Accessed 21 December 2015].

[34] Executive Office of the President. *The President's Climate Action Plan*. P. 10.

[35] United States EPA. *U.S. Greenhouse Gas Emissions* in 2013. Available from: http://www3.epa.gov/climatechange/ghgemissions/gases.html [Accessed 21 December 2015].

[36] US EPA. Oil and Natural Gas Sector: *Emissions Standards for New and Modified Sources, Proposed Rule*. Washington. 40 CFR Part 60. 2015.

[37] Environmental Protection Agency. *Emission Guidelines, Compliance Times, and Standards of Performance for Municipal Solid Waste Landfills; Proposed Rules*. Washington. 40 CFR Part 60. 2015.

[38] Executive Office of the President. *The President's Climate Action Plan*. P. 10.

[39] United States EPA. *U.S. Greenhouse Gas Emissions* in 2013. Available from: http://www3.epa.gov/climatechange/ghgemissions/gases.html [Accessed 21 December 2015].

[40] US Department of State. *United States Climate Action Report 2014, First Biennial Report of the United States of America, Sixth National Communication of the United States of America, Under the United Nations Framework Convention on Climate Change*. 2014. P. 11.

[41] The list of rules is available from: http://www.epa.gov/snap/snap-regulations [Accessed 21 December 2015].

[42] US Department of State. *Summary: North American 2015 HFC Submission to the Montreal Protocol*. Available from: http://www.state.gov/documents/organization/240964.pdf [Accessed 8 January 2016].

President Obama's CAP consists of 75 policies. The centerpiece is comprised of the regulations targeting power and transportation sector GHG emissions and other modest policies to improve energy efficiency, increase the use of renewable energy, and reduce non-CO_2 gases, while achieving other important policy objectives. An important omission is a strategy to tackle industrial emissions, which emit approximately 20 % of US GHG emissions.[43] What should we make of this approach? How does it compare with the path that is required to make progress at home in reducing GHG emissions?

For the most part, the policies embodied in the CAP are less ambitious, complex, and controversial than the BTU tax and ACES' cap-and-trade proposal. Because of this, it is unlikely they will be subjected to the same level of political attack. The CPP is the exception. The CAP policies attempt to hit a lot of singles and doubles and are contributing to the achievement of US emissions goals. The Administration has estimated that it is on track to meet its initial target of reducing GHG emissions 17 % below 2005 levels by 2020 and progress is being made to achieving the more ambitious target established for 2025.[44]

I believe the policies included in the US plan will receive greater support from the public and policy-makers as their GHG reductions and other co-benefits become more widely known. Assuming that comprehensive legislation is off the table for the foreseeable future, given continued disagreement on climate policy, many of the programs included in the plan can be ramped up to achieve additional reductions and other benefits. And although legislation is preferable to regulation for several reasons, one advantage of regulation is that the Executive Branch exerts greater control over rulemaking than legislation.

---

[43] US EPA. Sources of Greenhouse Gas Emissions. Available from: http://www3.epa.gov/climatechange/ghgemissions/sources.html [Accessed 21 December 2015].

[44] It is important to note that the US estimates that policies implemented and planned would reduce GHG emissions 22–27 % below 2005 levels by 2025 which is lower than the 26–28 % goal established for this year. The planned policies include proposed rules which have not yet been completed. For a description of these issues see Chapter four of the *United States Climate Action Report 2016.*

US Subnational Policies

Chapter 3 detailed some of the ongoing policy efforts by States in the US to reduce GHG emissions. A summary of some of the important initiatives follows.

1. *Emissions Targets*—As of August 2013, 29 states had some form of emission reduction targets.[45]
2. *Carbon Markets*—Approximately 10 states, including California and the nine Regional Greenhouse Gas Initiative (RGGI) states, have developed carbon markets.[46] There is the probability that this number will increase as states develop their plans to comply with the GHG emissions targets included in the CPP.
3. *Power Plant Standards*—As of February 2013, four states including California, New York, Washington, and Oregon imposed some sort of GHG emission standards on power plants.[47]
4. *Renewable Portfolio Standards*—As of 2015, 37 states required their power generators to supply their customers with electricity generated from a specific percentage of renewables or had goals in place.[48]
5. *Energy Efficiency Standards*—As of 2015, 23 states had policies in place requiring a specified percentage of energy be saved per year.[49]

These actions are making a contribution to GHG emissions mitigation in the US, while also creating political support for subnational action. The RGGI and California programs have shown that carbon markets are viable alternatives to cutting GHG emissions at the regional and state levels, potentially providing an impetus for other states to join such initiatives or to develop their own markets. And state renewable

---

[45] US Department of State. *United States Climate Action Report 2014.* P. 129.
[46] US Department of State. *United States Climate Action Report 2014.* PP. 128–129.
[47] US Department of State. *United States Climate Action Report 2014.* P. 129.
[48] US Department of State. *United States Climate Action Report 2016.* P. 28.
[49] US Department of State. *United States Climate Action Report 2016.* P. 28.

portfolio standards (RPS) are contributing to increased levels of renewables in the US.[50]

Some will argue that subnational policies are piecemeal and inefficient. However, the policies are providing important benefits such as building confidence and strengthening the institutions required to implement them.[51]

The International Arena

In addition to the domestic policies designed to achieve national GHG emission reduction targets, President Obama's CAP addresses adaptation and US efforts in the international arena. Adaptation is beyond the scope of this book. On the international front, the plan focused on the need to increase cooperation with key developing countries like China, India, and Brazil. It described progress made in the president's first term and the need to find new areas of cooperation during his second term.[52] The plan also focused on the progress made in the international negotiations in the president's first term and to seek a post-2020 agreement 'that is ambitious, inclusive and flexible'.[53] These efforts and their impacts are described later in this chapter.

International Climate Change Policy 2.0

This section describes the effort to develop a successor agreement to the KP and the emergence of international climate change policy 2.0. It elaborates the key decisions taken in the international negotiations that led to the Paris Agreement, the context in which the negotiations took place,

---

[50] Data indicated that RPS in place in 2008 would lead to the addition of 60 GW of renewables by 2025 which is equal to 15 % of electricity growth. See Plumer, B. (2008). 'A New Leaf'. Audubon Magazine, Volume 110(5), 62–73.

[51] Engel, K. (2006) 'State and Local Climate Change Initiatives: What is Motivating State and Local Governments to Address a Global Problem and What Does this Say About Federalism and Environmental Law', *The Urban Lawyer*, Vol. 38, (4), 1–17.

[52] Executive Office of the President. *The President's Climate Action Plan*. P. 17.

[53] Executive Office of the President. *The President's Climate Action Plan*. P. 21.

some of the key elements of the agreement, and the status and impacts of countries' pledges to reduce GHG emissions. I also provide general views on the agreement and the extent to which it conforms to the criteria that has put been forward to guide future policy-making efforts.

Key Events Leading Up to COP-21

Important decisions were taken in several COP meetings that culminated in the Paris agreement. A brief description of some of them follows.

Bali Action Plan

COP-13 was held in 2007 in Bali, prior to the initiation of the Kyoto commitment period. The meeting was held in the backdrop of the IPCC's Fourth assessment report that concluded 'that warming of the climate system is unequivocal'[54] and continued realization that deep cuts in GHG emissions were required to achieve the UNFCCC's long-term stabilization objective. In response, the COP '[d]ecides to launch a comprehensive process to enable the full, effective and sustained implementation of the Convention through long-term cooperative action, now, up to and beyond 2012, in order to reach an agreed outcome and adopt a decision at its fifteenth session, by addressing, inter alia'[55] mitigation by developed and developing countries, adaptation, technology development and transfer, and the provision of financial resources and investment for mitigation, adaptation, and technology cooperation.

For those not entirely familiar deciphering the exact meaning of the words included in these decisions, it signaled that the international community would attempt to reach a successor to the KP at COP-15 and was hopeful that a new President of the US would play a more constructive role in the process.

---

[54] IPCC, 2007: Climate Change 2007: Synthesis Report. Contribution of Working Groups I, II, and III to the Fourth Assessment Report of the Intergovernmental Panel on Climate Change (Core Writing Team, Pachauri, R.K. and Reisinger, A. [eds.]). IPCC, Geneva, Switzerland, 104 pp.

[55] *Report of the Conference of the Parties on its thirteenth session, held in Bali from 3 to 15 December 2007, Bali Action Plan.* Decision 1/CP.13. United Nations. 2008

The Copenhagen Accord

Two years after the Bali Action Plan was agreed on, Barack Obama had replaced George W. Bush as the US president. It is generally agreed that the administration's efforts in the international negotiations got off to a rocky start. The negotiations in Copenhagen did not achieve the objective established by the Bali Action Plan to adopt a decision on a post-2012 agreement. The disagreements that had bogged down the international climate change negotiations for years, particularly regarding developed and developing countries' future responsibilities to reduce GHG emissions, continued. This was the case even though the world had changed dramatically since the framework convention was agreed to in 1992 with respect to emissions growth and development. Developing countries' GHG emissions already surpassed developed countries' emissions[56] and are expected to account for 75% of emissions growth in the next 25 years.[57] Regarding development, nearly 50 non-Annex I (developing) countries have higher per capital incomes than the poorest Annex I countries (developed) and 40 non-Annex I countries ranked higher in the development index than the lowest ranked Annex I country.[58] Yet, in Copenhagen, it was like these changes had not occurred. Most viewed COP-15 as a failure and many believed that the ongoing international negotiating process conducted by the UN had run its course.

President Obama and several developing countries brokered a last-minute deal that became known as the Copenhagen Accord.[59, 60] Among

[56] This appears to be the case regarding CO_2 emissions. See **IEA (2011), CO_2 Emissions From Fuel Combustion 2011,** www.iea.org/statistics © OECD/IEA, Paris, IEA Publishing. Licence: www.iea.org/t&c.

[57] Leal-Arcas., R. *Top-down and Bottom-up Approaches in Climate Change and International Trade*: X ANNUAL CONFERENCE OF THE EURO-LATIN STUDY NETWORK ON INTEGRATION AND TRADE (ELSNIT) TRADE AND CLIMATE CHANGE, 2012, 19–20 October 2012, Milan, Italy. Arcas cites the data from US Energy Information Administration. *International Energy Outlook 2007*. US Department of Energy. 2015. Chapter 7. It was not available on the Internet.

[58] Leal-Arcas., R. *Top-down and Bottom-up Approaches in Climate Change and International Trade*. Arcas cites the data from the UN Development Programme. International Human Development Indicators. It was not available on the Internet.

[59] *Report of the Conference of the Parties on its fifteenth session, held in Copenhagen from 7 to 19 December 2009, The Copenhagen Accord*. Decision 2. CP 15. 2009.

[60] In an interview with Jeffrey Goodell in the Rolling Stone edition 8 October 2015, President Obama described the dysfunction in Copenhagen when he arrived and the difficulty in completing

other things, it agreed: (i) to limit temperature increases; (ii) to assist developing countries adapt to climate change and provide resources to do so; (iii) for developed countries to submit economy-wide emissions targets for 2020 and for developing countries to implement mitigation actions and to submit them; (iv) that developed countries would provide US\$ 30 billion to developing countries from 2010 to 2012 for mitigation, adaptation, technology development and transfer, and capacity building; and (v) committed to a goal of mobilizing US\$ 100 billion a year by 2020 to meet the needs of developing countries. A green climate fund was established to support these activities.

There was general hostility expressed toward the accord and the process for developing it. Most countries were cut out of the process, including the EU, which many view as the world's leader in fashioning a response to climate change. The full COP did not endorse the accord, although more than 100 countries agreed to it. The language in the accord states, 'the Conference of the Parties Takes note of the Copenhagen Accord 18 December 2009'.[61]

Although Copenhagen is frequently referred as a low point in the international negotiations, some believe that the debate began chipping away at the distinction between developed and developing countries with respect to GHG emissions mitigation, the key barrier to reducing global GHG emissions and to securing US participation in a global effort. Progress was made in subsequent COPs on this issue and it is the most important element of the new agreement reached in Paris at COP-21.[62]

Durban Platform for Enhanced Action

COP-17 was held in Durban, South Africa in 2011. Recognizing the gap between developed and developing countries' mitigation pledges and achieving the global communities' objective of limiting temperature increases to specified levels, the COP agreed to 'launch a process to

---

the Copenhagen Accord that ensured the participation of key countries including China and India in future efforts.

[61] *Report of the Conference of the Parties.* Copenhagen Accord. United Nations. 2009.

[62] *The Paris Agreement.* United Nations. 2015.

develop a protocol, another legal instrument or an agreed outcome with legal force under the Convention applicable to all Parties.'[63] The Parties also agreed that the work would be completed as early as possible, but no later than 2015, so it could be adopted at COP-21 in 2015 and come into force in 2020. The work would focus on mitigation, adaptation, finance, technology development and transfer, transparency of action, support and capacity building, and be informed by the IPCC's Fifth Assessment Report. Finally, a work plan on mitigation would be developed to close the ambition gap by ensuring the highest possible mitigation efforts by all Parties.[64]

Warsaw and Lima

At COP-19 in Warsaw, to advance what was agreed to in Durban, Parties were invited 'to initiate or intensify domestic preparations for their intended nationally determined contributions (INDCs)' in preparation for COP-21.[65] At COP-20 in Lima, Parties were invited to submit their INDCs well in advance of COP-21, by the first quarter of 2015 for those ready to do so. Guidance was given to Parties regarding the information that could accompany the INDCs to facilitate their 'clarity, transparency, and understanding'.[66]

Intended Nationally Determined Contributions

An INDC represents a country's commitment to GHG emissions mitigation in the post-2020 time frame, when a successor to the KP was scheduled to take effect. It represents a new, bottom-up approach to mitigation.

---

[63] *Report of the Conference of the Parties on its seventeenth session, held in Durban from 28 November to 11 December 2011, Establishment of an Ad Hoc Working Group on the Durban Platform for Enhanced Action.* Decision 1/CP.17. United Nations. 2012.

[64] *Report of the Conference of the Parties.* Durban Platform for Enhanced Action. United Nations. 2012.

[65] *Report of the Conference of the Parties on its nineteenth session, held in Warsaw from 11 to 23 November 2013, Further advancing the Durban Platform.* Decision 1/CP.19. United Nations. 2014.

[66] *Report of the Conference of the Parties on its twentieth session, held in Lima from 1 December to 14 December 2014, Lima Call for Climate Action.* Decision 1/CP.20. United Nations. 2015

The INDC includes a country's GHG mitigation pledge and the policies it determines best able to achieve it. They are based on a country's unique national circumstances, priorities, and policy-making traditions. The KP's top-down approach to mitigation was entirely opposite. It included inflexible GHG emissions limitations for Annex I countries that were negotiated by many countries and which failed to consider each nation's conditions such as economic growth, emissions trends, and resource mix. And although each country could use preferred approaches to achieve their emission reduction commitments, the KP incented the use of market mechanisms for compliance.

The Context of Paris

There was a shift in approach taking place from international climate change 1.0, characterized by the top-down, inflexible KP, to climate change 2.0. A more flexible, modest, and bottom-up approach was emerging. This represented a potential sea change in the international policy architecture. The international negotiations described above which culminated in the Paris Agreement at COP-21 were taking place alongside a flurry of other ongoing activities. Some of these, which were reached outside of the formal negotiations, included bilateral and multilateral agreements among governments and others between governments and non-state actors such as international institutions, NGOs, and other members of civil society. These arrangements, which many call carbon clubs, had, and will continue to have, a significant impact on the ongoing multilateral negotiations. This approach, represented by important bilateral climate change agreements, has been a key element of President Obama's strategy to forge a successor to the KP. Some of these, which affected the negotiations, are the subject of discussion.

These governance models have long been a staple of international diplomacy and cooperation in many areas of economic, energy, and environmental policy. Researchers point to bilateral and multilateral trade arrangements as examples of this approach.[67] Regarding climate change, a

---

[67] For some background on these topics, see R. Leal-Arcas, *Top-down and Bottom-up Approaches in Climate Change and International Trade* (2012) and D. Victor, *Global Warming Gridlock* (2011).

driver of these arrangements is the urgency to find solutions and frustration with the top-down, multilateral UNFCCC process that resulted in the failed KP and which nearly collapsed after Copenhagen. These agreements, which can take many forms, will be prominent in climate change 2.0 and have been characterized as bottom-up, hybrid models comprised of both top-down and bottom-up elements, carbon clubs, experimental government, and building blocks to name a few.[68] Because no one approach is likely to dominate future policy-making as the UNFCCC's top-down approach has previously, it is not necessary to describe each one in detail. It is likely that the initiatives being undertaken in this period of experimentation will continue to operate alongside of and/or potentially be integrated into the ongoing UNFCCC process.

What follows is a general description of some of the approaches and their attraction.

Bilateral and Multilateral Initiatives/Carbon Clubs

These arrangements can be between governments, international institutions, or members of civil society such as NGOs and business, or a mix of the three. They are entered into by members that have common interest in finding solutions to climate change and commit to undertake specific activities. Their substantive focus can be broad, addressing such issues as mitigation, clean energy, technology cooperation, adaptation, and climate finance. Or they can be narrower, established to achieve a more limited goal such as reducing short-lived climate pollutants. The clubs' substantive emphasis and membership can be expanded over time. Given the limited size of their membership, decisions can be made expeditiously, cooperatively, and flexibly. This is in contrast to the experience with the international process, which has been time-consuming, contentious, and inflexible, requiring near unanimity to make any decisions.

---

[68] See Arcas and Victor above. Also see Andresen, S. *International Climate Negotiations: Top Down, Bottom-up or a Combination of Both?*, Abbott, K. *The Transnational Regime Complex for Climate Change*, Sabel, C. and D. Victor, *Governing Global Problems Under Uncertainty: Making Bottom-up Climate Change Policy Work*, Falkner, R., H. Stephan, and J. Vogler, International Climate Policy after Copenhagen: Towards a 'Building Blocks Approach, and D. Bodansky and E. Diringer, *Alternative Models for 2015 Climate Change Agreement.*

The current distribution of global GHG emissions makes such arrangements logical. Ten countries were responsible for approximately 66% of global CO_2 emissions in 2013.[69] If agreements can be fashioned between these nations/regions, or a few of them, on such issues as emissions mitigation, it would appear that significant progress could be made and that it could influence the actions of others. The US successfully used this strategy to give momentum to the Paris negotiations.

US and the Club Approach

Earlier in this chapter, I mentioned that President Obama's CAP committed to active engagement in the international arena, both on the bilateral front and in the international negotiations.

Limiting GHG emissions in the US and making any progress on climate change was always going to be challenging and controversial. It would be difficult, if not impossible to do so if developing countries were not actively participating in the effort. A central tenet of President Obama's climate strategy, as articulated in his CAP, was to increase bilateral cooperation and strengthen relationships with key developing countries. In the run up to Paris, the administration built on previous cooperation and struck agreements with China, India, and Brazil. These countries have exerted great influence in the international negotiations and frequently been at odds with the US on how best to address climate change. Importantly, they are also the world's largest, and fourth and seventh largest GHG emitters in the world, respectively. It is not an exaggeration to conclude that these agreements represented the administrations' greatest impacts on climate change on the international stage prior to and in the run up to COP-21.

Together, the three agreements reached between the US and China, India, and Brazil cover approximately 50% of global GHG emissions. Although they would be important at any time, one of the US goals, among many, were to provide positive momentum to the Paris talks

---

[69] IEA (2015), CO_2 Emissions From Fuel Combustion 2015, **www.iea.org/statistics** © OECD/ IEA, Paris, IEA Publishing. Licence: www.iea.org/t&c P. 20.

by encouraging a growing spirit of cooperation and for other nations to increase the ambition of their commitments. The US–China agreement was particularly important to the Paris talks, given that the commitments made to reduce GHG emissions were the most ambitious by either country to date. Furthermore, it signaled a new era of cooperation on climate change between the two nations that had been at significant odds on the issue. It is not possible to address climate change without the US and China.

The agreements illustrate the nations' common interests in working together. For the US, they provide an opportunity to deepen cooperation on all aspects of climate change with major developing country emitters, to potentially expand opportunities for US firms to tap into some of the largest energy markets in the world, and to broaden relationships with important countries. The partnerships allow China, India, and Brazil to potentially access US financial resources, and technological expertise, and to strengthen relationships with private US firms to achieve their climate change and energy-related goals.

A brief description of the agreements, which were announced within seven months in 2014 and 2015, follows.

(1) *US–China Agreements*—The US–China agreements reached in 2014 and 2015 are noteworthy given that these countries are the world's two largest emitters of GHGs, accounting for more than 40 % of global CO_2 emissions in 2013,[70] and they had previously been hesitant to commit to reducing their GHG emissions. In the first agreement, the US committed to reduce its net GHG emissions by 26 to 28 % by 2025 and make best efforts to achieve the more ambitious goal. China announced it would attempt to peak its emissions by 2030 and attempt to do so earlier.[71] Both of these

---

[70] IEA (2015), CO_2 Emissions From Fuel Combustion 2015, **www.iea.org/statistics** © OECD/ IEA, Paris, IEA Publishing. Licence: www.iea.org/t&c

[71] The White House Office of the Press Secretary. *US-China Joint Announcement on Climate Change.* Available from: https://www.whitehouse.gov/the-press-office/2014/11/11/us-china-joint-announcement-climate-change [Accessed 21 December 2015].

commitments were incorporated in each nation's INDC.[72] China also committed to attempt to increase its share of non-fossil energy to 20 % by 2030.[73]

In the context of the second agreement, China announced that it would initiate a national cap-and-trade system in 2017 covering power generation and key industrial sectors.[74] Not many people would have bet that China would develop a market-based system for reducing GHG emissions. The US finalized the CPP prior to the second agreement and also made other commitments.[75]

The agreements also communicated the countries' vision for the Paris agreement, to reduce HFC emissions and to cooperate in the areas of climate finance, technological advancement, clean energy, and adaptation.[76]

(2) *US–India Agreement*—Building on prior cooperation, the US and India reached an agreement on Climate and Clean Energy Cooperation. This is another important agreement, given that India is the world's fourth largest emitter, accounting for nearly 7 % of global GHG emissions[77] and its influence is increasing in the international negotiations. As is described in Chap. 7, India's emissions will continue to grow as its economy expands. The agreement deepens the countries cooperation on climate change, including negotiations on COP-21, HFCs, clean energy, and climate finance.[78]

---

[72] United States of America. *US Intended Nationally Determined Contribution.*

[73] The White House Office of the Press Secretary. *US-China Joint Announcement on Climate Change.*

[74] The White House Office of the Press Secretary. U.S.-China Joint Presidential Statement on Climate Change.

[75] The White House Office of the Press Secretary. *U.S.-China Joint Presidential Statement on Climate Change.*

[76] See the 2014 and 2015 US–China agreements cited above.

[77] Friedrich, J. M. Ge, T. Damassa. Infographic: *What Do Your Country's Emissions Look Like?* Available from: http://www.wri.org/blog/2015/06/infographic-what-do-your-countrys-emissions-look [Accessed December 22 2015].

[78] The White House Office of the Press Secretary. *Fact Sheet: US and India Climate and Clean Energy Cooperation.* Available from: https://www.whitehouse.gov/the-press-office/2015/01/25/fact-sheet-us-and-india-climate-and-clean-energy-cooperation [Accessed 21 December 2015].

Although the agreement did not address mitigation, India committed to reduce its GHG emissions intensity by 33–35 % by 2030 from 2005 levels in its INDC and for non-fossil energy to account for 40 % of its power generation in the same year.[79]

(3) *US–Brazil Agreement*—The US and Brazil issued a joint statement on climate change. Brazil is currently the seventh largest emitter of GHGs in the world, accounting for more than 2 % of the global total.[80] The agreement puts forward a common vision of the key elements of the Paris Agreement including INDCs, updating of mitigation commitments, and long-term strategies for transitioning to a clean energy economy. Brazil also committed to meeting 28–33 % of its energy needs through non-hydro renewables. The US and Brazil also committed to cooperate to phase down HFCs, adaptation, and clean energy while managing forests, land use, and agriculture. These activities will be coordinated and managed by a new high-level climate change working group.[81]

Brazil committed in its INDC to reduce its GHG emissions 37 % below 2005 levels by 2025 and 43 % below those levels by 2030.[82]

In addition to the big three bilateral agreements, the Obama Administration also played a lead role in establishing the international arrangements that follow.

(4) *Major Economies Forum on Energy and Climate*—President Obama launched the forum which is comprised of 17 nations that account for 75 % of global GHG emissions to support the international

---

[79] India. *India's Intended Nationally Determined Contribution.* Available from: http://www4.unfccc. int/submissions/INDC/Published%20Documents/India/1/INDIA%20INDC%20TO%20 UNFCCC.pdf [Accessed 21 December 2015].

[80] Friedrich, J., M. Ge, T. Damassa. Infographic: What Do Your Country's Emissions Look Like?

[81] The White House Office of the Press Secretary. *US-Brazil Joint Statement on Climate Change 30 June 2015.* Available from: https://www.whitehouse.gov/the-press-office/2015/06/30/us-brazil-joint-statement-climate-change [Accessed 21 December 2015].

[82] Federal Republic of Brazil. *Brazil's Intended Nationally Determined Contribution.* Available from: http://www4.unfccc.int/submissions/INDC/Published%20Documents/Brazil/1/BRAZIL%20 iNDC%20english%20FINAL.pdf [Accessed 21 December 2015].

climate change negotiations and to focus on clean energy and energy efficiency.[83] President George Bush held a meeting of the forum in 2007.

(5) *The Climate and Clean Air Coalition to Address Short-Lived Climate Pollutants*—The Administration established the Coalition to Reduce Short-Lived Climate Pollution in 2012. It has expanded to 110 participants at the time of this writing including 50 countries, and a combined 60 international governmental and non-governmental organizations. The coalition's current emphasis is to implement 10 initiatives to reduce short-lived pollutants in landfills, the oil and gas sector, cook stoves, and diesel engines.[84]

The Administration also described a number of other ongoing multilateral initiatives including ones designed to expand trade in environmental goods and services, and reduce emissions from deforestation and degradation.[85]

These examples are solely within a US context. There is no shortage of existing organizations/clubs in the world that already have been or could be created to address climate-related issues. It is likely that they will continue to exert influence on climate change policy for the foreseeable future. Given the multitude of sub national and national trading programs in operation and under development around the world, the conditions exist to create trading clubs.

Some believe these arrangements have a greater chance of achieving their objectives than those reached by diplomats in the context of the multilateral climate change negotiations that resulted in the KP. If negotiations stall internationally, they could supplant the existing international negotiations and/or operate in conjunction with them.

---

[83] Executive Office of the President. *The President's Climate Action Plan.*

[84] A description of the Coalitions mission, a list of participants and initiatives can be seen on its website. Available from: http://www.ccacoalition.org/en [Accessed 21 December 2015].

[85] Executive Office of the President. *President Obama's Climate Action Plan, 2nd Anniversary Progress Report*, PP. 19–22.

The Potential Promise and Pitfalls of the Club Approach

And because bilateral and multilateral arrangements/clubs are likely to continue to exert their influence on climate change policy, I wanted to briefly review a few of the potential benefits they can provide and challenges they will confront.

Some Potential Benefits of the Club Approach

Some potential benefits from these arrangements follow.

(1) The formation of a club signifies member's common interests. This alone makes it easier to agree on an agenda and to achieve results than in multilateral negotiations with nearly 200 countries. Achievements reached within clubs can potentially influence the agenda within the larger UNFCCC process and positively impact its outcomes. The US bilateral agreements are examples of how this has occurred within the context of the Paris negotiations.

(2) Decision-making within clubs should be more expeditious and less cumbersome than within the context of the international negotiations.

(3) Based on the recent US agreements, it appears that clubs may be able to achieve important breakthroughs on mitigation, which is also in contrast from the experience with the UNFCCC progress. Because the large majority of global GHG emissions are concentrated within a small group of nations, agreements among several of them could go a long way to achieving climate policy objectives.

Some Potential Concerns with the Club Approach

Some concerns with these arrangements follow.

(1) Because like-minded nations would be the clubs' members, there is the potential that they would be hesitant to making ambitious mitigation commitments. One of the benefits of the international process

is that its prominence and public nature can be an incentive for members to increase their ambitions. Of course, the flip side is that some of the goals that have been established to date have been unrealistic.

(2) An important question within clubs is how compliance would be enforced. How would the club members sanction non-compliance? Would there be penalties? If so, would there be institutions to do so?

(3) By their nature, clubs exclude others that are not members. This can breed resentment and reduce the potential for cooperation between members and non-club members in other forums. For example, clubs comprised of big emitters would exclude those smaller nations that may be most affected by the impacts of climate change. These smaller nations typically prefer the protection and voice they are granted within the UNFCCC process to advocate for their interests.

The Paris Agreement

The Paris Agreement, which was reached at COP-21, has just been completed at the time of writing this. This section describes some of its key elements, status, and impacts of the INDCs in achieving long-term climate policy objectives, what they and some of the other provisions included in the agreement mean for climate protection, and other important issues that will require resolution in subsequent negotiations.

Similar to the differences in approach taken between climate change 1.0 and 2.0 in the US, the agreement reached at COP-21 was in stark contrast to the KP. Its centerpiece is a flexible, bottom-up approach to reducing GHG emissions that is combined with elements of a top-down regime.[86]

Increased Ambition

The agreement increased the ambition of climate change policy from the goal of limiting temperature increases to 2 °C to 'holding the increase in the global average temperature to well below 2 °C above preindustrial

---

[86] *The Paris Agreement to the United Nations Framework Convention on Climate Change.* United Nations. 2015.

levels and pursuing efforts to limit the temperature increase to 1.5 °C above pre-industrial levels'.[87]

To achieve the objective of limiting temperature increases, the agreement calls for a rapid reduction in GHG emissions after emission peak, 'so as to achieve a balance between anthropogenic emissions by sources and removals by sinks of greenhouse gases in the second half of this century'.[88] Some are interpreting this provision to mean that GHG emissions should be net zero by 2050.

Nationally Determined Contributions (NDCs) and Supporting Provisions

NDCs are the foundation of the Paris Agreement with respect to mitigation. (This term has replaced INDCs in the Paris Agreement and will be referred to as [I]NDCs for the remainder of the book). Each party is required to 'prepare, communicate and maintain successive nationally determined contributions'.[89] Parties' successive [I]NDCs, are required every five years[90] and to reflect the Parties 'highest possible ambition'.[91] Developed countries' [I]NDCs will include economy-wide absolute emission reduction targets, while developing countries should continue their mitigation efforts and are encouraged to move toward economy-wide targets.[92] The sum total of countries' GHG emissions reduction commitments included in the [I]NDCs will determine the shape of global GHG emissions pathways for the post-2020 period and the subsequent efforts that will be required throughout the twenty-first century to achieve the international community's goal of limiting temperature increases to well below 2 °C.

Several top-down provisions in the agreement are designed to ensure the [I]NDCs' transparency and integrity and whether Parties are making

---

[87] *Paris Agreement.* Article 2 1. (a).
[88] *Paris Agreement.* Article 4 1.
[89] *Paris Agreement.* Article 4 2.
[90] *Paris Agreement.* Article 4 9.
[91] *Paris Agreement.* Article 4 3.
[92] *Paris Agreement.* Article 4 4.

progress in achieving their targets. A brief review of such provisions follows. Parties are required to 'provide the information necessary for clarity, transparency and understanding' when they communicate their [I] NDCs.[93] They are also required to account for their [I]NDCs. In doing so, 'Parties shall promote environmental integrity, transparency, accuracy, completeness, comparability and consistency, and ensure the avoidance of double counting'.[94]

The agreement also includes an enhanced transparency framework.[95] One of its primary purposes is to track Parties' progress in achieving their [I]NDCs.[96] To do so, Parties are required to provide a national inventory of GHG emissions and removal by sinks and 'information necessary to track progress made in implementing and achieving its nationally determined contribution'.[97] The information 'will undergo a technical expert review'[98] and include a consideration regarding the implementation and achievement of Parties' NDC and areas for improvement.[99]

The agreement also calls for the COP to 'take stock of the implementation of the agreement to assess the collective progress towards achieving the purpose of this Agreement and its long-term goals (the global stocktake)'.[100] This first assessment would take place in 2023 and be undertaken every five years.[101] The results would be used by Parties in updating their [I]NDCs.[102]

The Agreement also establishes a mechanism to promote compliance.[103]

---

[93] *Paris Agreement.* Article 4 8.
[94] *Paris Agreement.* Article 4 13.
[95] *Paris Agreement.* Article 13.
[96] *Paris Agreement.* Article 13 5.
[97] *Paris Agreement.* Article 13 7.
[98] *Paris Agreement.* Article 13 11.
[99] *Paris Agreement.* Article 13 12.
[100] *Paris Agreement.* Article 14 1.
[101] *Paris Agreement.* Article 14 2.
[102] *Paris Agreement.* Article 14 3.
[103] *Paris Agreement.* Article 15.

Market Mechanism

An article devoted to markets is included in the agreement. Although its intention is somewhat unclear given the lack of detail, it appears to have two objectives.

A new mechanism was established to contribute to the mitigation of GHG emissions and support sustainable development.[104] There appears to be important distinctions between it and the CDM and JI. An important difference is that it seeks to move beyond offsetting emissions 'to deliver an overall mitigation in global emissions'.[105] This requirement could serve as an incentive for developing countries to adopt absolute GHG emissions targets to trade with developed countries. Another important distinction with the CDM and JI is that there appears to be no limitations on Parties' participation. The CDM authorized transfers of CERs between Annex I and Non-Annex I countries and participation in JI was limited to Annex I countries. By taking this approach, the mechanism seeks to minimize the distinction between developing and developed countries, similar to the [I]NDCs. The COP is tasked with developing the rules to govern the mechanism.[106]

The second objective appears to be encouraging the linkage of trading systems and carbon markets by allowing for the use of 'internationally transferred mitigation outcomes' to comply with [I]NDC targets. In other words, countries can trade their reductions with others which can then use them to comply with [I]NDC targets. In doing so, Parties are required to 'promote sustainable development and ensure environmental integrity…and put in place a robust accounting framework to guard against double counting and which is consistent with guidance provided by the COP'.[107] Because there are many carbon markets in operation or under development, this could go a long way toward encouraging their

---

[104] *Paris Agreement.* Article 6.4.
[105] *Paris Agreement.* Article 6.4.(d).
[106] *Paris Agreement.* Article 6.7.
[107] *Paris Agreement.* Article 6.2.

linkage and potentially the creation of larger markets. The COP is tasked with developing the rules to govern the mechanism.[108]

Other

Numerous provisions are included in the agreement for developed countries to provide financial and technical support to developing countries to support their efforts.

The Impacts of the [I]NDCs to Date

As of 15 December 2015, 160 [I]NDCs, representing 187 countries and covering nearly 99 % of global GHG emissions, had been submitted to the UNFCCC secretariat.[109] The elements included in the [I]NDCs illustrate the strengths and weaknesses of the approach. The strengths are that each country determines its target based on their circumstances and the policies to achieve them. The weakness is the [I]NDCs lack of standardization. A UNFCCC synthesis report of the [I]NDCs indicates this challenge. They include several different types of GHG targets (absolute, relative, and intensity), different base years to measure them and different end points. Some include long-term targets and others do not. They cover different sectors and GHGs. Parties also used different approaches to account for their land use activities. Some condition parts of their implementation on the receipt of assistance from other countries.[110] Importantly, the lack of [I]NDCs standardization makes it more difficult to secure the benefits of carbon markets and to link them.

The most important question regarding the [I]NDCs is whether they can be successful in achieving the international community's climate

---

[108] *Paris Agreement.* Article 6.2.

[109] World Resources Institute. *CAIT Climate Data Explorer: Paris Contributions Map.* Available from: http://cait.wri.org/indc/ [Accessed 22 December 2015].

[110] For a review of the components of Parties [I]NDCs, see Section I. C. of the UNFCCCs, *Synthesis report on the aggregate effect of the intended nationally determined contributions.* 2015.

policy objectives. Three analyses undertaken of the [I]NDCs which had been submitted in preparation for COP-21 are unanimous in their conclusion that they would not be adequate to limit temperature increases to 2 °C which had been the objective of climate policy prior to Paris.

United Nations Framework Convention on Climate Change INDC Synthesis Report

The UNFCCCs synthesis report assessed the impact of 119 [I]NDCs communicated by 147 Parties by 1 October 2015, which covered 80 % of global GHG emissions.[111] All scenarios showed higher emissions in 2025 and 2030 compared with 1990, 2000, and 2010 base years.[112] The [I]NDCs do result in lower emissions than in the pre-[I]NDC scenarios.[113] The analysis shows that emissions levels based on the [I]NDCs assessed for the report are estimated to be nearly nine $GtCO_2e$ higher in 2025 and 15 $GtCO_2e$ higher in 2030 than are necessary to achieve the two-degree target in a least-cost scenario.[114]

PBL Netherlands Environmental Assessment Agency

The PBL Netherlands Environmental Assessment Agency assessed the impacts of 186 [I]NDCs submitted as of 29 November 2015, covering approximately 96 % of 2012 global GHG emissions. It concluded that implementation of the [I]NDCs would achieve reductions between 9 and 11 $GtCO_2e$, leaving a gap of between 12 and 14 $GtCO_2e$ to achieve the objective.[115]

---

[111] United Nations Framework Convention on Climate Change. *Synthesis report on the aggregate effect of the intended nationally determined contributions.* United Nations. FCCC/CP/2015/7. 2015. P. 4.

[112] United Nations Framework Convention on Climate Change. *Synthesis report.* PP. 40–41.

[113] United Nations Framework Convention on Climate Change. *Synthesis report.* P. 43.

[114] United Nations Framework Convention on Climate Change. *Synthesis report.* P. 47.

[115] PBL Netherlands Environmental Assessment Agency. *PBL Climate Pledge INDC Tool.* Available from: http://infographics.pbl.nl/indc/ [Accessed 22 December 2015].

Climate Action Tracker

Analysis undertaken by the Climate Action Tracker assessed the impacts of 158 [I]NDCs, representing 185 countries and covering 94% of global GHG emissions. The analysis, which includes many assumptions regarding the post-2030 period, found that if the [I]NDCs were fully implemented in 2025 and 2030, temperature increases would be limited to 2.7 °C by 2100. However, a gap of 11–13 $GtCO_2e$ in 2025 and 15–17 $GtCO_2e$ in 2030 from current pledges would need to be closed for the world to get on an emissions pathway consistent with the prior two-degree target.[116]

Greater reductions in GHG emissions will be required to limit temperature increases to well below two degrees Celsius than what was assumed by the analysis cited above because it was based on limiting increases to 2 °C. However, I view the trends cited above with some measure of optimism for a few reasons. Importantly, the Paris Agreement sends a clear and consistent message to the industry that the international community is taking climate change seriously and that this framework could potentially be in place for decades. In addition, this is the first time the [I]NDC approach has been used to date. I would expect that a country's ambition would increase as more knowledge is gained regarding the policies that are have the greatest success in reducing GHG emissions and technological options expand as their performance improves and costs come down.

Carbon Markets in the [I]NDCs

Carbon markets were the preferred policy instrument in climate change 1.0. So how did they fare in the [I]NDCs? As of January 2016, approximately 53% of the submitted [I]NDCs indicated that they planned to or it would be possible that they would use international market mechanisms, 35% did not specify, while slightly more than 11% would not

---

[116] Climate Action Tracker. *2.7°C is not enough – we can get lower*. Available from: http://climateactiontracker.org/assets/publications/briefing_papers/CAT_Temp_Update_COP21.pdf [Accessed 22 December 2015].

use the mechanisms to comply with their targets.[117] We know the US and the EU has already stated that they will not rely on international market mechanisms to achieve their GHG emissions targets.[118]

The Potential Benefits and Pitfalls of the [I]NDC Approach

This chapter has communicated the differences in the approaches of the top-down KP and the bottom-up Paris Agreement. Because of the failure of the KP to slow the growth in GHG emissions, a new approach was urgently needed. It is not possible at this time to know whether [I]NDCs will be successful in achieving the international community's goal of limiting temperature increases. Some of the strengths and weaknesses of the approach are highlighted below.

Potential Benefits of the [I]NDC Approach

A few of the potential strengths of the [I]NDC approach follow.

(1) By selecting their own targets, countries have a vested interest in achieving them. Failure would be an embarrassment.

(2) The [I]NDCs are flexible, encourage experimentation, and provide the ability to scale-up successful policies throughout the twenty-first century. They should allow countries to learn and adapt to new information.

(3) Each country's [I]NDC's will logically be structured to deal with the component parts of its national GHG emissions. For example, approximately 60 % of US emissions result from the combustion of fossil fuels for power generation and transportation. This is why the two most ambitious components of the US strategy are regulations targeting the GHG emissions from these sectors. Other policies have

---

[117] World Resources Institute. *CAIT Climate Data Explorer: Paris Contributions Map.*

[118] The US and EU INDCs are available from: http://www4.unfccc.int/submissions/INDC/Submission%20Pages/submissions.aspx [Accessed 22 December 2015].

also been put in place to improve energy and carbon intensity and to strategically targeting non-CO_2 gases, which comprise nearly 20 % of US GHG emissions.

The [I]NDC approach is consistent with the strategies for climate change 2.0 that has been described. It is more modest and flexible than the approach to mitigation taken in 1.0. Although they hold promise, analysis of the initial commitments indicates that much more ambition will be required in successive [I]NDCs.

It will also be important to define [I]NDCs' success in a realistic fashion. Most will judge them solely on the basis of whether they achieved the objective to limit temperature increases to the agreed on level. That is fair. However, they should not be evaluated solely on that basis. Other metrics for evaluation include the reductions that they achieve and the climate impacts they help prevent.

Potential Pitfalls of the [I]NDC Approach

(1) There is the probability that [I]NDCs may represent a race to the bottom. Because each nation has both an incentive to achieve their targets and not to impose higher cost on its economy, it could respond by adopting overly modest targets. This warrants watching.

(2) [I]NDCs need to become more standardized. As mentioned previously, the [I]NDCs submitted to date differ in many important respects. Although this diversity is a natural outgrowth of the initial process and recognition of countries' different circumstances, the lack of standardization will make it difficult for countries to engage in cost-saving trade and create efficient carbon markets, compare countries' level of effort, and to evaluate the progress that is being made toward the achievement of long-term climate policy objectives. Standardization would address these challenges.

(3) The [I]NDCs have the potential to reduce economic efficiency and increase fragmentation. This is because as countries move forward with their own carbon pricing policies, the use of potentially incompatible approaches and design elements can further complicate an

already complex process for linking markets. Although this is a natural outgrowth of the [I]NDC approach, it can adversely affect the ability to limit costs, which remains an important goal of climate change policy.

The goal in identifying some of the potential benefits and pitfalls of the [I]NDC process and carbon clubs is to illustrate that like all approaches, they are not a panacea. The weaknesses inherent in these approaches can hopefully be addressed by top-down provisions, which were incorporated in the Paris Agreement, and as experience is gained.

Observations About the Paris Agreement

My general views on the Paris Agreements approach to mitigation follow.

(1) I like the agreements' universality requiring [I]NDCs from both developed and developing countries, which is in contrast to the KP. An agreement would not have been possible without such an approach.
(2) I believe for the reasons cited above that the bottom-up approach to mitigation represented by [I]NDCs can be successful.
(3) I am hopeful that the top-down elements included in the agreement are designed to ensure rigor and discipline in the development of the [I]NDCs. Furthermore, the transparency framework incorporating an expert review of them could increase the potential for countries to live up to their commitments.
(4) The global stocktake provision (another top-down element), assessing the collective progress made in achieving the long-term goals of the agreement, may serve as an incentive for Parties to increase the ambitions of successive NDCs.
(5) I am skeptical as to whether a new market mechanism can be successful, but I like its emphasis on achieving reductions in contrast to only offsetting emissions and the attempt to encourage widespread participation.

(6) I am also encouraged by the authorization of international transfers of mitigation outcomes, which could stimulate linkage of trading systems and the creation of larger carbon markets. This could help control the costs of mitigation and serve as an incentive for increased ambition in successive [I]NDCs.

(7) I believe the agreement has a chance to succeed because it attempts to create a consistent, enduring framework to address climate change throughout the twenty-first century. This is essential to provide the private sector with the signals and incentives to make the necessary investments to decarbonize the energy system.

Although there are many elements of the Paris agreement to like, and the negotiators who developed it should be applauded, it remains conceptual. It will not work unless effective rules are put in place to implement it.

The first five chapters of the book reviewed the events and policies that dominated climate change 1.0 in the US and at the international level. This chapter reviewed the policies being put in place in the transition to climate change policy 2.0 and contrasted them with the prior efforts. The failures of the past 25 years have served as a catalyst to the new direction being taken. The defeat of ACES was a wake-up call to what could realistically be achieved in the US. Similarly, the demise of the KP, the failure of Copenhagen, and the near collapse of the international negotiating process may have had a positive effect. Processes sometimes need to fail or collapse before they can be put back together. Hopefully, we will be able to look back in 25 years from now and conclude that these new models began to put global GHG emissions on a sustainable pathway. If this occurs, the prior failures will not have been for naught.

7

Recommendations

I have attempted to articulate the lessons that can be drawn from the US and international efforts to advance climate change policy in the first generation of policy-making and reviewed the direction of new efforts in Chaps. 1, 2, 3, 4, 5, and 6. This chapter focuses on the future.

The Criteria Underpinning the Recommendations

I have provided my views gained from 25 years in government and business to communicate the reasons why the policies proposed and adopted in 1.0 failed when measured by growing GHG emissions, rising concentrations, and the need to start over. In this chapter, I focus on applying those lessons by making recommendations to guide future policy-making. There are six broad recommendations in this section. The first four are applicable to both the US and the international community; the last two are focused predominantly on the US.

© The Author(s) 2016
R.H. Rosenzweig, *Global Climate Change Policy and Carbon Markets*, Energy, Climate and the Environment,
DOI 10.1057/978-1-137-56051-3_7

The Need to Start Modestly

My criteria to guide the development of climate change policies, and which influenced the recommendations in this chapter, follow. Policies need to set realistic targets, be modest, and be flexible. A policy is more likely to be successful when it starts modestly, is targeted to achieve specific objectives, and aims at building a foundation than when it starts big. This approach can result in significant reductions in GHG emissions.

My views regarding policies' future emphasis are reflected in an article on state policy that quotes Terry Tamminen, who served as the former Environmental Secretary in California, and Chief Policy Adviser to former Governor Schwarzenegger. The discussion that follows explains states' efforts in developing responses to climate change and some of the reasons for their success. 'Although this seems like a piecemeal approach to a colossal problem, that's part of why some states have succeeded. Tamminen observes that one reason that states have not been afflicted by the legislative paralysis that's plaguing Congress is that they do not try to ram through one big climate bill all at once. 'If you look at the wedges'—referring to the different sectors of the economy that can be decarbonized—'You tackle each one individually, with six or eight or ten bills', says Tamminen. In Florida, Crist (referring to the former governor) was unable to pass a comprehensive bill similar to California's Global Warming Solutions Act through the legislature, so instead he passed an energy bill promoting renewables and efficiency. 'If you score the greenhouse-gas benefits of that bill, he's halfway to the goal', Tamminen notes. 'Sure, it's sleight of hand, but that's what it takes to get it done'. These smaller measures add up: Nationwide state renewable energy standards will force utilities to add 60 gigawatts of renewable capacity by 2025, equal to 15% of the projected growth in electricity, according to the Lawrence Berkeley National Laboratory. And the Center for Climate Strategies has estimated that if all 50 states simply adopted policies similar to those in place in 12 'leadership' states—such as Arizona, North Carolina, and New York—the US could slash emissions by 33% by 2020 and save about US$25 billion in the process'.[1]

---

[1] Plummer, B. A New Leaf, *Audubon Magazine*, Volume 110(5) September–October 2008. P. 4.

The analysis may or may not be correct. But I agree with the approach that is advocated. Modest policies add up. A lot of singles and doubles. They can be successful in reducing GHG emissions and achieve other objectives. What would have happened if the US Congress took such an approach in 2009–2010 and then attempted to build on success? Maybe started with a utility only cap-and-trade bill, or a modest energy bill? We will never know.

Many will criticize the path I am suggesting and as described above to be inadequate, given the magnitude of the climate change problem, and destined to fail. I believe the opposite. Policy-makers need to avoid the temptation to try and hit home runs. We have been down that path before with ACES and the KP and it did not work. Unfortunately, the current fragility and fractured nature that currently characterizes our political system does not allow for great ambition.

If policy-makers are going to be able to fashion the more ambitious solutions that will be required to achieve long-term climate policy objectives, they will first need to earn the public's trust, which does not exist today. Achieving results is the only way to do so. I believe that modest policies designed to achieve clearly defined objectives can succeed, creating the conditions and the consensus necessary for the more ambitious actions that will be necessary in the future.

For the most part, the recommendations that follow are modest in their ambition, targeted to achieve specific objectives, provide co-benefits, and control costs. The large majority of them are designed to cut CO_2 emissions from the energy sector, which remains by far the largest single contributor to the GHG emissions that contribute to climate change. And the recommendations are general, because many of them have been the focus of lengthy analysis and in the public domain for several years.

The issue of cost control, which dominated 1.0, must remain a focus. There can be a political backlash if policies' costs rise to levels determined to be unacceptable and unnecessary to aid industries capable of competing with lower levels of support. This has resulted in reductions of renewables subsidies in Germany, the UK, and Spain.[2]

---

[2] Witte, G. Britain pulls the plug on renewable energy. The Washington Post. 21 November 2015.

The word modest is used here to characterize the policies' breadth and goals. However, some of the measures could be considered ambitious given the levels of reductions they could achieve and the difficulties they will confront in the political process. But their scope and intent are relatively narrow.

The Magnitude of the Climate Challenge

The challenge in successfully confronting climate change is beyond question. Although many human activities contribute to the GHG emissions that cause climate change, CO_2 emissions accounted for nearly 75% (primarily from energy production and use and to a lesser extent agriculture) of global GHG emissions in 2010.[3] The energy sector is responsible for approximately 66% of total anthropogenic GHG emissions,[4] with CO_2 emissions from fossil fuel combustion accounting for more than 90% of this amount.[5]

Achieving the international community's goals of limiting temperature increases to well below 2 °C cannot be achieved without dramatically reducing CO_2 emissions from the energy sector. These emissions persist in the atmosphere for a century or longer and have the greatest impact on the accumulation of CO_2e concentrations. It is generally accepted that an equivalent 450 ppmv CO_2e concentration level provides a two-thirds probability of holding temperature increases below 2 ° pre-industrial times.[6, 7]

---

[3] US Environmental Protection Agency. *Global Greenhouse Gas Emissions.* Available from: http://www3.epa.gov/climatechange/science/indicators/ghg/global-ghg-emissions.html [Accessed 22 December 2015].

[4] ©OECD/IEA 2015, World Energy Outlook Special Report on Energy and Climate, IEA Publishing. Licence: https://www.iea.org/t&c/. P. 20.

[5] ©OECD/IEA 2015, World Energy Outlook Special Report on Energy and Climate, IEA Publishing. Licence: https://www.iea.org/t&c/. P. 25.

[6] IPCC, 2014: Summary for Policymakers. In: Climate Change 2014: Mitigation of Climate Change. Contribution of Working Group III to the Fifth Assessment Report of the Intergovernmental Panel on Climate Change (Edenhofer, O., R. Pichs-Madruga, Y. Sokona, E. Farahani, S. Kadner, K. Seyboth, A. Adler, I. Baum, S. Brunner, P. Eickemeier, B. Kriemann, J. Savolainen, S. Schlömer, C. von Stechow, T. Zwickel and J.C. Minx [eds.]). Cambridge University Press, Cambridge, United Kingdom and New York, NY, USA. P. 10.

[7] For the most part, this book was completed before COP-21 which increased the ambition of climate change policy to holding temperature increases to well below 2 °C from the prior goal of 2 °C. The IPCCs AR 5 analysis was based on the 2 degree target and that is why it is referenced in this section. It is logical to assume that a concentration ceiling lower than 450 PPMV will be required to achieve the more ambitious target.

The world has already passed the 400 ppmv level and it is adding approximately 2 ppmv per year.[8]

What will it take to get there? Achieving the 450 ppmv CO_2e ceiling will require that emissions be reduced from current levels by 40% to 70% by 2050 and be near zero or negative by 2100.[9] To accomplish this, the amount of low carbon emitting energy supplies will need to increase from current levels by approximately 145% in 2030 and 310% globally in 2050.[10, 11]

GHG emissions are determined by four variables. These include population, per capita GDP, energy intensity of economic activity, and carbon intensity of energy.[12] Population is expected to grow from 7.3 billion to 9.7 billion by 2050,[13] and the global economy, if it grows by 3.5% until 2040 as estimated by the International Energy Agency, will be two and a half times bigger than it is currently.[14]

As two of the variables affecting GHG emissions are expected to grow significantly, dramatic improvements in energy and carbon intensity are the only way to provide energy to fuel economic growth, provide electricity to more than one billion people in the world who currently lack access to it, and to hold temperature increases below 2 °C. To make these improvements, policies will need to lower the costs of low- and non-emitting energy while increasing the costs of fossil fuels in a politically acceptable fashion. The recommendations in this chapter are designed with this goal in mind.

---

[8] Dr. Pieter Tans, NOAA/ESRL (www.esrl.noaa.gov/gmd/ccgg/trends/) and Dr. Ralph Keeling, Scripps Institution of Oceanography (scrippsco2.ucsd.edu/).

[9] IPCC, 2014: Summary for Policymakers. In: Climate Change 2014: Mitigation of Climate Change. Contribution of Working Group III to the Fifth Assessment Report of the Intergovernmental Panel on Climate Change. PP. 10–12.

[10] IPCC, 2014: Summary for Policymakers. In: Climate Change 2014: Mitigation of Climate Change. Contribution of Working Group III to the Fifth Assessment Report of the Intergovernmental Panel on Climate Change. P. 9.

[11] Achieving the new target will required greater levels of emission reductions and the addition of more low and non-emitting energy.

[12] The equation that follows was developed by Yoichi Kaya who currently serves as President of the Research Institute of Innovative Technology for the Earth has been used to calculate GHG emissions. Total emissions = population × (GDP/population) × (energy/GDP) × (emissions/energy).

[13] United Nations, Department of Economic and Social Affairs, Population Division. World Population Prospects: The 2015 Revision, Key Findings and Advance Tables. Working Paper No. ESA/P/WP.241. 2015.

[14] ©OECD/IEA 2015, World Energy Outlook Special Report on Energy and Climate, IEA Publishing. Licence: https://www.iea.org/t&c/. P. 33.

Progress Made to Date in Climate Change 2.0 and Challenges

In Chap. 6, I cited the current status of the [I]NDC submissions and described the results of three analyses regarding their impacts. Although some progress is being made to reduce GHG emissions, the improvements in reducing carbon intensity and improving energy efficiency are inadequate to keep pace with population and economic growth.

The challenge in determining progress and what needs to be done in the future requires a review of both recent trends and future projections. Snapshots of both follow. The good news first. Regarding decarbonization trends, renewables accounted for nearly 50% of new power supplies in 2014.[15] And the energy intensity of the global economy declined by more than 2% in 2014, double the rate of the prior decade. This translated into a stabilization of energy-related CO_2 emissions in 2014, while the global economy grew at 3%. Declines occurred in the EU and Japan, with a modest increase in the US.[16]

As for future projections, continued progress in decarbonization and improvements in energy intensity is forecast to continue through the 2040 period. More than half of the new power supply added by 2040 is expected to be renewables.[17] Non-fossil sources increase to 25% of the energy mix in 2040 from the current level of 19%.[18] And renewables for power generation reach 50% in the EU, approximately 30% in China and Japan, and 25% in the US and India.[19] This results in coal use for power generation declining from 41% to 30% globally and to less than 15% outside of Asia.[20] Energy efficiency policies proliferate around the

---

[15] ©OECD/IEA 2015, World Energy Outlook Special Report on Energy and Climate, IEA Publishing. Licence: https://www.iea.org/t&c/. P. 11.

[16] ©OECD/IEA 2015, World Energy Outlook Special Report on Energy and Climate, IEA Publishing. Licence: https://www.iea.org/t&c/. P. 29.

[17] ©OECD/IEA 2015, World Energy Outlook, IEA Publishing. Licence: https://www.iea.org/t&c/ P. 5.

[18] ©OECD/IEA 2015, World Energy Outlook, IEA Publishing. Licence: https://www.iea.org/t&c/ P. 1.

[19] ©OECD/IEA 2015, World Energy Outlook, IEA Publishing. Licence: https://www.iea.org/t&c/ P. 6.

[20] ©OECD/IEA 2015, World Energy Outlook, IEA Publishing. Licence: https://www.iea.org/t&c/ P. 6.

world limiting demand growth to 30% till 2040, while the global economy grows by 150%,[21] significantly reducing energy intensity per unit of GDP. Although this is good news, it is not nearly enough to address the climate challenge.

In short, much work needs to be done for this progress to take place. It cannot be assumed. Economic and technological trends often do not occur as planned, and governments do not always live up to their commitments. And the analysis of the initial [I]NDCs cited in Chap. 6 indicates that current commitments are inadequate to limit temperature increases to the levels agreed to by the international community.

Geographical Considerations

The context of addressing the climate challenge has changed dramatically since the UNFCCC was agreed. Non-OECD energy-related CO_2 emissions have over taken those of the OECD. They accounted for approximately 60% of global emissions in 2014, with China alone accounting for almost half of that amount while India's share is growing.[22] China and India's needs for energy indicate that these trends will continue.

China

China has made significant energy- and climate-related commitments in recent years. These include the deployment of a large amount of non-fossil energy, increasing fuel economy standards in transportation, improving energy efficiency, and to introduce an ETS by 2017. The IEA indicates that these policies combined with structural changes in its economy lead to a peak in its emissions by 2030.[23] Even if China achieves its objectives, it is host to 50% of the world's fleet of coal-fired power plants used by

---

[21] ©OECD/IEA 2015, World Energy Outlook, IEA Publishing. Licence: https://www.iea.org/t&c/ P. 6.

[22] ©OECD/IEA 2015, World Energy Outlook Special Report on Energy and Climate, IEA Publishing. Licence: https://www.iea.org/t&c/. PP. 27–28.

[23] ©OECD/IEA 2015, World Energy Outlook, IEA Publishing. Licence: https://www.iea.org/t&c/ PP. 1–2.

2030[24] and becomes the world's largest oil user by 2030.[25] Its energy-related CO_2 emissions are two and a half times the US levels in 2030, which remains the world's second largest emitter.[26]

India

India's need to provide electricity to nearly 300 million people who currently lack access, and to support economic growth will have a major impact on its energy use and GHG emissions. If India achieves its goals for GHG emissions and to increase its amount of non-fossil energy included as included in its [I]NDC, [27] coal will still account for nearly 50% of its energy supply and it will become the world's largest source of growth in coal use and the largest importer.[28] Oil demand also increases more than that in any other country by 2040.[29] An analysis of India's commitments prior to the submission of its [I]NDC but inclusive of many of its goals indicates that its energy-related CO_2 emissions increase by approximately 60% from 2013 to 2030.[30]

Long-term goals for global GHG emissions can only be achieved by making substantial improvements in these countries and throughout Asia.

---

[24] ©OECD/IEA 2015, World Energy Outlook Special Report on Energy and Climate, IEA Publishing. Licence: https://www.iea.org/t&c/ P. 52.

[25] ©OECD/IEA 2015, World Energy Outlook Special Report on Energy and Climate, IEA Publishing. Licence: https://www.iea.org/t&c/ P. 52.

[26] ©OECD/IEA 2015, World Energy Outlook Special Report on Energy and Climate, IEA Publishing. Licence: https://www.iea.org/t&c/ P. 51.

[27] India. *India's Intended Nationally Determined Contribution.* Available from: http://www4.unfccc. int/submissions/INDC/Published%20Documents/India/1/INDIA%20INDC%20TO%20 UNFCCC.pdf [Accessed 21 December 2015].

[28] ©OECD/IEA 2015, World Energy Outlook, IEA Publishing. Licence: https://www.iea.org/t&c/ P. 2.

[29] ©OECD/IEA 2015, World Energy Outlook, IEA Publishing. Licence: https://www.iea.org/t&c/ P. 2.

[30] ©OECD/IEA 2015, World Energy Outlook Special Report on Energy and Climate, IEA Publishing. Licence: https://www.iea.org/t&c/ P. 55.

Recommendations

This remainder of this chapter makes broad-based recommendations. They focus on next steps regarding implementation of the Paris Agreement, carbon markets under development, and the need to gradually increase the cost of fossil fuels while creating a robust portfolio of low- and non-emitting energy technologies necessary to improve carbon and energy intensity and to control costs.

Some of these recommendations are applicable to all governments. Others are focused solely on the US because of my knowledge and experience in national energy and environmental policy and the political system. I am aware that many question whether the US will stay the course on climate change chartered by President Obama, given the ideological and partisan differences that continue to exist. My assumption in developing the recommendations is that it will.

In addition, one area the recommendations are silent on is 'carbon pricing'. Carbon prices will be created in the 60 jurisdictions that have either implemented or are in the process of developing tax or trading programs. I also assume this trend will continue. But as discussed throughout, this bottom-up approach is entirely different from what was envisioned decades ago. I cannot add much to the millions of words that have already been written on the subject. Because of my policy and market experience, I suggest my most useful contribution in this area was the analysis of the EU ETS and CDM performance in Chap. 4 and the general recommendations for carbon market design in this chapter. And although the US will not have a national price on carbon, signals will be sent through the subnational systems already in place and the rules targeting the transportation and power sectors, which account for approximately 60 % of US emissions.

Recommendation 1: Get the Rules in Place to Implement the Paris Agreement

The Paris Agreement represents either a turning point or another failed approach. I provided my general views of some of the provisions in the

Paris Agreement in Chap. 6. It holds promise, but is only the beginning. The provisions included in the agreement remain concepts at this time. Bringing them to life and achieving the promise of Paris requires tough, persistent bargaining and climate diplomacy to put the necessary rules in place to implement the agreement in an expeditious fashion. Many significant questions remain. I will raise a few to illustrate the arduous nature of the task ahead.

Based on my experience, I emphasize the [I]NDCs, the new market mechanism, and guidance for countries to link trading systems. My views on the key areas for attention follow.

Nationally Determined Contributions

Rules are required to guide the [I]NDCs' development, increase their standardization, ensure their transparency, and to assess whether they have achieved their targets. In addition, the global stocktake, the process put in place to assess 'collective' progress made in achieving the agreements long-term goals and to inform the ambition of successive [I]NDCs, should also be a priority.

Standardization

As described in Chap. 6, a glaring weakness of the initial [I]NDCs is their lack of standardization. The group tasked with preparing for the Paris Agreement's entry into force should provide the Parties with the necessary guidance on issues to improve the [I]NDCs in this regard. At a minimum, guidance is required in the areas that follow: (i) base year, (ii) target year, (iii) coverage, (iv) moving toward common types of targets, (v) methodologies for accounting for emissions and removals by sinks, and, (vi) submissions within common time frames. If this does not occur, expanding cooperative approaches to reducing GHG emissions, linking carbon markets, comparing levels of effort, and evaluating progress will be difficult.

Transparency

Guidance should be provided to Parties regarding the information that will 'facilitate clarity, transparency, and understanding of nationally determined contributions'.[31]

Accounting

Clear rules that provide sufficient clarity to the Parties regarding the accounting for emissions and sequestration in their [I]NDCs, and consistency in the communication and implementation of their [I]NDC are required.[32] This is necessary to achieve several important objectives in successive [I]NDCs, including environmental integrity, transparency, and comparability.[33]

Assessment

Rapid action is required to develop the rules that will enable the implementation of the 'enhanced transparency framework for action and support'. This is an important provision of the agreement designed to enable an expert review and assessment of developed countries' support of developing countries and its implementation and achievement of its [I]NDC.

Putting the flesh on the bones of this concept will be difficult. A few critical issues that need to be addressed follow. Who will undertake the expert review? How frequently will it occur? What information will the Parties be required to provide so that its progress in achieving its [I]NDC can be assessed? How will the review recommend areas of improvement for the Parties? These are but a few issues that require resolution to develop an effective process.

---

[31] *Paris Agreement to the United Nations Framework Convention on Climate Change.* Paragraph 28. United Nations. 2015.

[32] *Paris Agreement.* Paragraph 31.

[33] *Paris Agreement.* Article 4. 13.

Global Stocktake

This is another critical element of the agreement because its purpose is to determine the progress made in achieving the long-term goals of the agreement and to inform the development and ambition of Parties' successive [I]NDCs. Countries require a clear understanding of the progress being made to achieve the agreement's long-term objective, and the reductions in GHG emissions that will be required throughout the twenty-first century in preparation for increasing the ambition of successive [I]NDCs.

Once again, many issues must be addressed to implement this provision. Who will undertake the stocktake? What will be its nature? How will it be used to spur greater ambition? Will it be integrated into other ongoing assessments?

In short, the effective implementation of the Paris Agreement requires the top-down provisions included in the agreement be put in place to impose discipline on the preparation of [I]NDCs, ensure their transparency, and assess nations' progress in meeting their targets while evaluating the collective progress being made in achieving long-term objectives.

Market Mechanism

As evidenced by the inclusion of Article 6, market mechanisms remain a priority of governments and maintain the political support of a large part of the business community. Without business support, a continued commitment to markets would not have been made. Because of this, rules to govern the mechanism included in the Paris Agreement should be developed expeditiously, hopefully learning from the prior experience with the CDM and JI.

The mechanisms' emphasis should be to achieve larger-scale mitigation than the KP's project-based mechanisms allowed for and to achieve widespread participation. It appears that this is the intent of Article 6. Another goal should be to ease the administrative burden.

The [I]NDCs' lack of standardization, particularly with respect to different types of targets serves as an obstacle to the effective implementation of the market mechanism. For the most part, developed countries' targets are described as absolute and tonnage based. Tons are standardized units and can be traded, measured, and tracked with ease. In contrast, some key developing countries opted for different types of targets denominated as intensity- or rate-based or other. The lack of standardization in this regard discourages trade among countries with different types of targets. You cannot trade tons and rates effectively. And emissions can increase even as improvement in emissions intensity occurs, minimizing the potential for trade between these countries to achieve reductions. This one example illustrates the challenge in achieving the mechanisms' objective '[t]o deliver an overall mitigation in global emissions'.[34] This is why standardization is important and developing countries should be encouraged to move toward absolute targets.

In addition, many developing countries accounting frameworks for GHG emissions are in their nascent stages. This too can prevent cooperation because such systems are critical to ensuring environmental integrity. Because of this, developed countries should assist developing countries create the accounting infrastructure that is necessary to effectuate trade.

Enabling Linkage of Carbon Markets

In addition to the global mechanism, the agreement envisions the linkage of trading systems by allowing the transfers of 'internationally transferred mitigation outcomes' by Parties to meet [I]NDC targets. There are requirements for Parties that engage in these activities to take action 'consistent with guidance adopted by the COP'.[35] Because of the potential benefits that linkage can provide, the COP should move quickly to provide such guidance. Emphasis should be placed on assuring the environmental integrity of units that are transferred between Parties and accounting standards to guard against double-counting.

The lack of [I]NDCs' standardization will serve as an obstacle to effective linkage so it should be prioritized.

---

[34] *Paris Agreement.* Article 6. 4. (d).

[35] *Paris Agreement.* Article 6. 2.

Recommendation 2: Establish Mutually Beneficial Approaches to Working with China, India, and Others with Growing Emissions

China and India continue to be major players on the global climate stage because of rapidly increasing energy use to meet the needs of their populations. This creates the necessity and the opportunity for the developed world, and from my perspective, especially the US, to develop closer collaboration with China and India in the energy sphere. It is in the developed world's interest to share best practices and to collaborate in energy research development and demonstration (RD&D) to develop and deploy more efficient energy technologies to enable the transformation of their energy systems.

The framework for such collaboration could be within the context of the Paris Agreement, bilateral or multilateral clubs, or existing international institutions.

Recommendation 3: Design the Carbon Market Rules Right

The purpose of the carbon markets established by the KP and the EU ETS was to create a single carbon price around the globe. In theory, this would create an incentive to implement emission reduction activities where it was cheapest to do so. For reasons described throughout, this approach has run its course at the global level; although another attempt will be made through the mechanism authorized by the Paris Agreement.

The term 'carbon pricing' generally refers to some form of emissions trading program or energy tax. It replaced the term 'carbon markets' because of its decline in popularity based on experience with the KP mechanisms and the EU ETS. What follows is the current state of play on carbon pricing initiatives at the regional, national, and subnational levels and general recommendations for a path forward.

The EU ETS

The performance of the EU ETS was described in Chap. 4. We know there was an enormous oversupply of EUAs that began in Phase 2, created by the ETS and the combination of CPs, imports of project-based reductions, and the recession that continued into Phase 3 and drove down EUA prices. Typical of top-down systems, the EU was unable to respond to market conditions in a timely fashion, which prolonged the situation. It is unknown whether the Market Stability Reserve (MSR) that has been put in place will work to normalize supply and demand and increase EUA prices.

The EU has doubled down by developing a Phase 4 of the EU ETS that will run from 2021 to 2030, requiring covered sectors to reduce their emissions by 43% in 2030 from 2005 levels. For the market to function better, the EU will need to determine the MSR's effectiveness in normalizing supply and demand, and take timely actions if it is not. This is critical, given the performance of the EU ETS over the past several years. The EU will also need to design the EU ETS and CPs, such as the RED and EED, better to prevent the adverse impacts that their interactions created in Phases 2 and 3. They will need to monitor them.[36] This will be challenging, given that the ambition of these policies has been increased.

National and Subnational Markets

With the exception of the EU ETS, which is a partially top-down—albeit regional system, bottom-up national and subnational carbon pricing systems continue to be developed in climate change 2.0. About 39 nations either have implemented or are scheduled to implement a carbon pricing program. This consists of 21 nations that either have or are going to implement an emission trading scheme, four that have implemented or are scheduled to implement a carbon tax, and 14 that either have

---

[36] For a description of the interactions that can occur between an ETS or tax and other energy policies and ways to better integrate them, see Christina Hood, ©OECD/IEA 2013, Managing Interactions between carbon pricing and existing energy policies, IEA Publishing. Licence: https://www.iea.org/t&c.

implemented or are planning to implement both. And 23 subnational jurisdictions have a carbon pricing program in place, consisting of 22 trading programs including the seven Chinese pilot efforts and one tax.[37] These programs cover seven $GtCO_2e$ or 12 % of global GHG emissions, with two-thirds of emissions covered by an ETS and one-third covered by a carbon tax.[38] The EU ETS, Chinese pilot programs and US programs cover 3.5 $GtCO_2e$ or 50 % of the total.[39] The carbon prices in these systems range between US$1 and US$130 a ton of CO_2 and are valued at under US$50 billion in total.[40] Another analysis concludes that the carbon price within these systems is US$7 per ton of CO_2.[41] The big news recently has been China's statement that it intends to introduce a national ETS in 2017. Without rendering a judgment on this, we should wait to see how this plays itself out.

These systems represent bottom-up efforts when compared with the KP and EU ETS. Because of this, their design elements are different. My view is that they would be most effective by: (a) including declining caps over multiple periods to provide certainty to regulated firms and investors, and (b) limiting coverage to power sector emissions, although many of the systems either cover or contemplate covering the manufacturing sectors' emissions while some also cover transport. Out of concerns with competitiveness, governments have utilized various mechanisms in an attempt to protect energy-intensive manufacturing sectors from higher direct and indirect costs. There may be a less complicated way to regulate such industries including taxes and/or regulation. Transportation sector emissions can be covered by imposing fuel economy and GHG standards.

---

[37] Kossoy, A., G. Peszko, K.Oppermann, N. Prytz, N. Klein, K. Blok, L. Lam, L, Wong, Bram Borkent. 2015. State and Trends of Carbon Pricing 2015 (September), by World Bank, Washington, DC. P. 11.

[38] Kossoy, A., G. Peszko, K. Oppermann, N. Prytz, N. Klein, K. Blok, L. Lam, L. Wong, B. Borkent. 2015. State and Trends of Carbon Pricing 2015. P. 21.

[39] Kossoy, A., G. Peszko, K. Oppermann, N. Prytz, N. Klein, K. Blok, L. Lam, L. Wong, B. Borkent. 2015. State and Trends of Carbon Pricing 2015. P. 10.

[40] Kossoy, A., G. Peszko, K. Oppermann, N. Prytz, N. Klein, K. Blok, L. Lam, L. Wong, B. Borkent. 2015. State and Trends of Carbon Pricing 2015. P. 21.

[41] ©OECD/IEA 2015, World Energy Outlook Special Report on Energy and Climate, IEA Publishing. Licence: https://www.iea.org/t&c/ P. 23.

The most common concern of governments in designing such systems is the potential for high allowance prices and price volatility. Many, if not all, include mechanisms to guard against this including banking and some types of limited borrowing, allowance reserves, and the use of offsets to name a few.[42] My view is that banking and borrowing (with some limits) should always be permitted. Allowance reserves have the potential to operate like a tax and reduce the potential benefits a market can provide. Offset programs can provide important benefits including cost control, stimulating innovation in sectors not covered by the trading program, and serving as a bridge until better technology is available, enabling firms to achieve reductions within their own assets at lower costs. The biggest challenge with offset systems will be to develop approaches to effectively determine additionality. Governments should continue to experiment with positive lists, standardization, and other methods in the attempt to strike the careful balance between providing certainty to developers, and stimulating market development while maintaining environmental integrity.

For the reasons cited throughout the book, and based on my experience with the CDM, offset systems will confront challenges. They will be controversial and are unlikely to provide the levels of cost control that analysis usually assumes. However, I believe that there is a better likelihood that subnational and national offset systems will achieve their objectives better than global mechanisms for several reasons. The entities administering them have greater expertise than a UN organization and they will also have more efficient decision-making processes in place. Governments at these levels should be able to adapt to new information and resolve problems in a more timely fashion than hierarchal approaches allow. In short, my view is that US EPA, and other government agencies would administer an offset program more effectively than a UN bureaucracy.

In addition, the subnational and national policy-making process benefits from the input of stakeholders who are familiar with the policy-making traditions and circumstances unique to each jurisdiction. This increases the potential to achieve consensus and resolve problems.

---

[42] For a description of many of these programs see, The International Emissions Trading Association, CDC Climate Research, and Environmental Defense Fund, 'The World's Carbon Markets: A Case Study Guide to Emissions Trading', 2015.

Regardless of the challenges experienced with carbon markets to date, my view is that they can still enable cost-effective GHG emission reductions. Because of this they should have a place in the policy portfolio in climate change 2.0. Like all environmental markets, their performance will be uneven for the reasons cited throughout the book. However, there is the potential they will work better than prior programs because of lessons learned and the fact that market oversight and decision-making is vested in individual governments, which are more experienced working with interested stakeholders in a transparent fashion.

Policy-makers at the national and subnational levels utilizing carbon pricing systems must also consider their interaction with CPs. Governments at all levels are using both approaches to reduce GHG emissions and to achieve other objectives. The interaction of the EU ETS and CPs had far-reaching impacts. My view is that one approach should take precedence over the other based on the government's primary objectives. If the government is seeking to achieve many objectives and co-benefits, CPs may take precedence over markets as the leading policy approach with markets playing a more supportive role. If the primary objective is to control costs of achieving reductions in GHG emissions, the use of markets may be preferred as the dominant instrument.

Markets in the US

It is unlikely that the US will adopt a national emissions trading program for many years for the reasons cited throughout. However, calls for cap-and-trade may increase as regulated firms seek a more uniform approach than allowed for in the CPP. If this were to occur, my recommendation would be to limit it to the power sector.

Move Carefully to Get the Benefits of Linking

Linkage is receiving widespread attention because of the growth of national and subnational systems and the fractured nature of the international markets. In general, linking authorizes the trade of GHG compliance instruments between firms located in linked jurisdictions and mutual recognition for compliance. An example of linking is the state of California and the province of Quebec. Many have advocated for jurisdictions to link their trading systems to create larger, more liquid markets. The argument is that the coverage of a larger number of sources with different abatement costs in multiple jurisdictions provides regulated firms with greater opportunities for trade and to lower compliance costs. Some believe that the creation of a market through such a process, and outside of the UNFCCC, would be beneficial.[43] I generally agree with these views and where it makes sense, jurisdictions should link their markets to take advantage of cost-saving opportunities. However, because of the provision in the Paris Agreement for Parties participating in such activities to do so consistent with guidance provided by the COP,[44] those interested in linking will want to closely monitor international developments.

As with all things regarding carbon markets, the theory of linkage sounds right. However, linking of trading systems also raises complex economic, political, and design issues. One such issue that has been discussed previously is comparability. For example, if the linking party's targets are not of comparable ambition, this could provide an opportunity for firms in the jurisdiction with weaker targets to make money from the sale of compliance instruments to entities in the jurisdiction with a more ambitious target. This would raise political issues. There are many issues like this that would require resolution for linkage to succeed.[45] Because of the complexity of the issues involved and the time and resources required to link systems, my recommendation is for jurisdictions considering linking is to start slow. Moving too rapidly increases the chance of failure.[46, 47] In contrast, and consistent with my overall views regarding future policy, a more modest, limited initial approach that succeeds in the beginning could lead to a more ambitious effort in the long run.

---

[43] Green, J. F., T. Sterner, G. Wagner (2014) 'A balance of bottom-up and top-down in linking climate policies'. *Nature Climate Change*, 4, 1064–1067.

[44] *Paris Agreement.* Article 6. 2.

[45] Green, J.F., T. Sterner, G. Wagner (2014) 'A balance of bottom-up and top-down in linking climate policies'.

[46] Green, J. F., T. Sterner, G. Wagner (2014) 'A balance of bottom-up and top-down in linking climate policies'.

[47] Green, J. F. A realistic approach to linking carbon markets. Washington Post, 1 December 2014.

Recommendation 4: Gradually Eliminate Fossil Fuel Subsidies

All energy sources are subsidized, including the production and consumption of fossil fuels. The subsidies are in different forms and can be hard to define. No common definition of subsidy exists today. One analysis identifies seven different types of subsidies provided to both consumers and producers of fossil fuels.[48] The OECD considers subsidies to be direct budgetary transfers and tax expenditures that support the production or consumption of fossil fuels compared with alternatives.[49] It has compiled an inventory of nearly 800 measures provided by member nations and six partner countries (Brazil, China, Indonesia, India, Russia, and South Africa) that support the production and consumption of fossil fuels.[50] To illustrate how prevalent such subsidies are, in 2014, 13% of CO_2 emissions were linked to fossil fuel use encouraged by a subsidy valued at US\$115 per ton. This is compared with 11% of global energy emissions covered by a carbon price averaging US\$7 per ton.[51]

Types of Energy Subsidies

This section describes two types of fossil fuel subsidies.

Consumer Subsidies

In general, these are subsidies in which governments provide some type of support that result in energy consumers paying costs below a benchmark price.

---

[48] UNEP Green Economy Policy Brief. *Fossil Fuel Subsidies*. Available from: http://www.unep.org/greeneconomy/Portals/88/documents/GE_BriefFossilFuelSubsidies_EN_Web.pdf [Accessed 23 December 2015].

[49] OECD. *OECD Companion to the Inventory of Support Measures for Fossil Fuels 2015*, OECD Publishing, Paris. DOI: http://dx.doi.org/10.1787/9789264239616-en. 2015. P. 10.

[50] OECD. *OECD Companion 2015*. P. 10.

[51] ©OECD/IEA 2015, World Energy Outlook Special Report on Energy and Climate, IEA Publishing. Licence: https://www.iea.org/t&c/ P. 17.

Producer Subsidies

In general, these are subsidies in which governments provide some type of support to producers that increase their profitability from what it would be.

Impacts of Fossil Fuel Subsidies

The analysis concludes that fossil fuel subsidies have adverse economic, public health, and environmental impacts. These include the budgetary costs to finance the subsidies, potentially reducing resources for other priorities. The subsidies are typically regressive, even though they are often put in place to shield low-income groups from high energy prices.[52] And by keeping the costs of fossil fuels lower than they otherwise would be, subsidies increase consumption and local air pollution. From a climate perspective, the subsidies result in increased GHG emissions and potentially discourage investment in renewables and efficiency.[53]

Estimating the Value of Fossil Fuel Subsidies

Estimating the value of fossil fuel subsidies is extremely complex and is dependent on the methodology used. What follows are the results from four recent analyses, the first focuses on the OECD and six nations, the second and third focus on global subsidies, and the fourth on the US.

OECD

OECD estimates that the nearly 800 budget and tax measures documented in its inventory of fossil fuel subsidies in member countries and

---

[52] Data on the regressive nature of the subsidies can be found in, International Monetary Fund, *Energy Subsidy Reform: Lessons and Implications*. 2013, and a Joint Report by IEA, OECD, OPEC and World Bank on fossil fuel and other energy subsidies: *An update of the G-20 and Pittsburgh and Toronto commitments*. 2011.

[53] The impacts of fossil fuel subsidies on local air pollution and climate change will be described later in this section.

six partner countries had a value of US$160 to US$200 billion annually from 2010 to 2014. Subsidies for the consumption of petroleum products constituted the 'bulk of that amount'.[54]

IEA

IEA estimates that global fossil fuel subsidies were nearly US$500 billion in 2014. This would have been more than US$600 billion had reforms not been enacted since 2009. In contrast, renewables received US$135 billion in subsidies in 2014, of which US$23 billion was for biofuels.[55] The IEA uses a price gap approach, which compares the prices end users pay for energy with a reference price.[56]

IMF

IMF estimates that the global value of fossil fuel subsidies was approximately US$5.6 trillion in 2015, representing 6.5% of the global GDP.[57] This is comprised of an estimate for pretax subsidies projected to reach US$330 billion in 2015 and post-tax subsidies of US$5.3 trillion in the same year.[58, 59] The estimate for post-tax subsidies reflects the underpayment for the environmental externalities resulting from energy consumption including air pollution, climate change, and estimates for foregone tax revenues.

---

[54] OECD. *OECD Companion 2015.* P. 10.

[55] ©OECD/IEA 2015, World Energy Outlook, IEA Publishing. Licence: https://www.iea.org/t&c/ P. 7.

[56] Coady, D., I. Parry, L. Sears, and B. Shang. *IMF Working Paper: How Large Are Global Energy Subsidies?* International Monetary Fund. WP/15/05. 2015. Appendix 1, P. 31.

[57] Coady, D., I. Parry, L. Sears, and B. Shang. *IMF Working Paper: How Large Are Global Energy Subsidies?*

[58] Coady, D., I. Parry, L. Sears, and B. Shang. *IMF Working Paper: How Large Are Global Energy Subsidies?* PP. 17–18.

[59] For a description of the data and methods used to arrive at the estimates of pre- and post-tax subsidies, see Coady, D., I. Parry, L. Sears and B. Shang in *How Large Are Global Energy Subsidies?* PP. 13–16.

The environmental externalities account for more than US$4.6 trillion of the US$5.3 trillion of the post-tax subsidies. Regarding the externalities created by specific energy sources, those created by coal use are highest, followed by petroleum and natural gas and electricity accounting for less. Local air pollution is estimated to be responsible for more than US$2.7 trillion, or nearly 60%, of the US$4.6 trillion of the post-tax subsidy attributed to externalities. Global warming accounts for more than 25%.[60] The post-tax subsidies are distributed across regions and their magnitude is dependent on fuel use.[61]

The US Analysis

Subsidies for fossil fuel production have a long tradition in the US. One recent analysis details approximately US$32.5 billion in subsidies in 2013 that were provided by the federal government and states for exploration, production, and consumption of oil, gas, and coal.[62] And similar to the IMF study, this analysis detailed US$350 to US$500 billion in externalities resulting from the combustion of fossil fuels in the US.[63]

Defining and estimating fossil fuel subsidies are a complex process.[64] However, a conclusion from the recent analyses above indicates that they are significant and represent a prominent element of national energy policies.

Benefits of Reducing Fossil Fuel Subsidies

Significant fiscal, environmental, and economic benefits would result from eliminating these subsidies. The IMF assumes the elimination of post-tax consumer subsidies would result in significant economic benefits

---

[60] These results can be found in Appendix 4, P. 37. of Coady, et al.

[61] Coady, D., I. Parry, L. Sears, and B. Shang. *IMF Working Paper: How Large Are Global Energy Subsidies?* PP. 20–22.

[62] Makhijani, S. *Cashing In On All Of The Above: U.S. Fossil Fuel Production Under Obama.* Oil Change International. 2014. The breakdown in exploration, production and consumption subsidies provided by the federal government and states can be found in Appendices I and II, PP. 17–23.

[63] Makhijani, S. *Cashing In On All Of The Above.* P. 15.

[64] For a description of how estimates cited were arrived at, see Appendix 1. P. 31. in D. Coady, I. Parry, L. Sears, and B. Shang in *How Large Are Global Energy Subsidies?*

and reduce CO_2 emissions by more than 20%. [65] Other analysis also concludes that the reduction or elimination of fossil fuel subsidies would cause a significant reduction in GHG emissions. The IEA argues that this is one of the four zero-cost actions that would keep the world on the trajectory to two degrees in the period to 2020.[66]

The Need to Move Slowly

Analysis indicates that reducing fossil fuel subsidies will create multiple benefits. In recognition of this, the leaders of the G-20 nations committed to 'rationalize and phase out over the medium term inefficient fossil fuel subsidies that encourage wasteful consumption'.[67] The G-20 has continued to call for this in subsequent summits. Leaders of the Asia Pacific Economic Cooperation made a similar statement in its 2009 Declaration.[68] Some progress has been made.[69]

The recommendation to phase out fossil fuel subsidies will be challenging given the entrenched interests in support of them and the importance of fossil fuels to the economies of many energy producing nations. There is a higher potential for success if phase out is attempted in an incremental fashion and is accompanied by companion measures that shield the poor from higher energy prices. It would also appear that the adverse public health impacts resulting from local air pollution, which comprise approximately 60% of the US$4.6 trillion in post-tax subsidies estimated by the IMF, would increase the potential for progress on this issue. This has influenced energy policies put in place by China. One analyst put forward a set of process-oriented recommendations that appears sensible and necessary to make progress in phasing out subsidies. These

---

[65] Coady, D., I. Parry, L. Sears, and B. Shang. *IMF Working Paper: How Large Are Global Energy Subsidies?* PP. 22–26.

[66] ©OECD/IEA 2014, Energy, Climate Change & Environment, IEA Publishing. Licence: https://www.iea.org/t&c P. 23.

[67] G-20. *Leaders Statement.* The Pittsburgh Summit. 2009.

[68] Asia Pacific Economic Cooperation. *Singapore Declaration: Sustaining Growth, Connecting the Region.* 2009.

[69] ©OECD/IEA 2015, World Energy Outlook Special Report on Energy and Climate, IEA Publishing. Licence: https://www.iea.org/t&c/ P. 17.

include developing a common international definition of subsidy, an agreed methodology to measure them, transparency in reporting them, and a peer review analysis for reforming them.[70] In the US context, it may be possible to create a diverse coalition of interests that would be interested in phasing out subsidies to achieve different objectives. Some on the right would have an interest in eliminating the so-called corporate welfare, while on some on the left would be interested in securing the environmental benefits that this policy could provide.

Recommendation 5: Increase the Development and Deployment of Low- and Non-Emitting Energy Technologies

The international community and the US require a robust, flexible portfolio of cost-competitive technologies to make the dramatic improvements in carbon and energy intensity necessary to reduce GHG emissions from the energy sector and to control costs.[71] Success requires an increased and sustained commitment to energy RD&D to improve the performance and lower the costs of existing technologies and to develop new ones, a strategy for investment, and other technology policies. Although many policies are necessary for the widespread development and deployment of energy technologies, it is generally accepted that investments in energy RD&D can make an important contribution in doing so. This is not a novel or original recommendation, but it may be the most important one. Unfortunately, spending on energy RD&D is more often than not influenced more by which political party is in charge, energy prices, and markets than by need.

---

[70] Whitley, S. *Time to change the game: Fossil fuel subsidies and climate.* Overseas Development Institute. PP. 21–22. 2013.

[71] For a description of the impacts that technology can have in controlling costs see, Edmonds, J., T. Wilson and R. Rosenzweig, *Global Energy Technology Strategy Addressing Climate Change: Initial Findings from an International Public-Private Collaboration.* Joint Global Change Research Institute. 2000.

Edmonds, J., J.J. Dooley, E.L. Malone, L.E. Clarke, S.H. Kim, J.P. Lurz, P.J. Runci, et al. *Global Energy, Technology and Climate Change Addressing Climate Change: Phase 2 Findings from an International Public-Private Sponsored Research Program.* Pacific Northwest National Laboratory, PNNL-SA-51712. 2007.

In addition to achieving technology innovation and climate benefits, RD&D can help achieve many other important objectives. The IEA concludes that government spending on energy research and development also results in improved productivity, job creation, expanding exports, and local environmental benefits.[72] A prominent group of US business leaders cite the benefits of increased innovation in energy including reduced risks of climate change, clean air, national security, and protection from energy price shocks.[73]

Increase Global RD&D Spending

Investment trends and gaps in clean energy technologies point to the need for increased spending on energy RD&D. Government investment in energy RD&D over the past three decades can be characterized as episodic and inadequate. From 1985 to 1995, government research in the nine OECD countries that undertook 96% of the energy RD&D at that time, declined by 23% in real terms.[74] Investments by the US fell by 23% from 1985 to 1998, and from 75% to 90% in Germany, Great Britain, and Italy from 1985 to 1995.[75]

There have been upticks in government investment in energy RD&D. Government spending increased from 1997, but except for the increases in 2009 included in stimulus programs as a response to the economic downturn, which represented a doubling from 2008 levels, spending had declined over the past 35 years in real terms.[76] And energy RD&D declined from 12% of the R&D total in 1981, to 4% in 2008.[77]

---

[72] ©OECD/IEA 2011, Clean Energy Progress report—IEA Input to the Clean Energy Ministerial, IEA Publishing. Licence: https://www.iea.org/t&c/. P. 32.

[73] The American Energy Innovation Council. *A Business Plan for America's Energy Future*. 2010. P. 4.

[74] Edmonds, J., T. Wilson, and R. Rosenzweig. *A Global Energy Technology Strategy Project Addressing Climate Change: An Initial Report of an International Public-Private Collaboration*. Joint Global Change Research Institute. 2000. P. 49.

[75] Edmonds, J., T. Wilson, and R. Rosenzweig. *A Global Energy Technology Strategy Project Addressing Climate Change*. Joint Global Change Research Institute. 2000. P. 50.

[76] ©OECD/IEA 2011, Clean Energy Progress report—IEA Input to the Clean Energy Ministerial, IEA Publishing. Licence: https://www.iea.org/t&c/. P. 6.

[77] ©OECD/IEA 2011, Clean Energy Progress report—IEA Input to the Clean Energy Ministerial, IEA Publishing. Licence: https://www.iea.org/t&c/. P. 6.

However, there have been significant increases in energy RD&D spending in China and India, which had previously not been major players in this area and other countries.[78]

A simple look at a government spending in energy RD&D is an imperfect metric of measuring progress and commitment to cleaner forms of energy. However, we do know that current efforts are inadequate given the estimated US$53 trillion in investment required in energy supply and efficiency until 2035 to bring the world on a path to two degrees or less.[79] We also know that government must play a unique role in energy innovation, given the resource constraints on the private sector, the magnitude of the required investment, and the inherent risks in RD&D.

There is a need for governments around the world to increase their investment in RD&D for clean energy technologies important to achieve climate policy and energy security objectives. An analysis of several technologies by the IEA concluded that an additional US$40 billion to US$90 billion per year is required to improve their cost competitiveness and performance from the US$10 billion spent by governments at the time of the analysis.[80, 81] Half of this amount or US$20 to US$45 billion would come from governments.[82]

Increase US RD&D Spending

The US needs to make increased investments in energy RD & D. Consider this: the US spends less than 0.05 % on energy RD&D as a percentage of national energy sales, which is less than national expenditures on potato

---

[78] Anadon, L. D., M. Bunn, G. Chan, M. Chan, C. Jones, R. Kempener, et al. *Transforming U.S. Energy Innovation*. Harvard Kennedy School, BELFER Center for Science and International Affairs. 2011. PP. 280–281.

[79] ©OECD/IEA 2014, World Energy Investment Outlook Special Report, IEA Publishing. Licence: https://www.iea.org/t&c/.

[80] ©OECD/IEA 2011, Clean Energy Progress report—IEA Input to the Clean Energy Ministerial, IEA Publishing. Licence: https://www.iea.org/t&c/. P. 14.

[81] These technologies are nuclear power, advanced vehicles, high efficiency coal units, bioenergy, solar, CCS, energy efficiency, smart grid, and wind. See P. 15 above.

[82] ©OECD/IEA 2011, Clean Energy Progress report—IEA Input to the Clean Energy Ministerial, IEA Publishing. Licence: https://www.iea.org/t&c/. P. 14.

and tortilla chips.[83] Energy R&D is less than 0.8% of the energy expenditure in the US, compared with the total 2.8% of R&D to the total economy.[84] Government investment in energy RD&D in the US is low by any measure.

A prominent group of business leaders recommended that the US increase its spending from US$5 billion in a typical year to US$16 billion, arguing that spending has declined for 30 years and was 25% of 1978 levels.[85] Similar recommendations have been put forward by high-level groups such as the President's Council of Advisors on Science and Technology, the National Commission on Energy Policy, the IEA, and the IPCC.[86] There may be a chance to increase RD&D spending in the US. Unlike many issues in the US political system, there is general agreement that government plays an important role in the process.

Put a Dedicated Mechanism in Place in the US to Fund Energy RD&D

To be successful, increases in energy RD&D need to be consistent and sustained. It currently is not. Energy RD&D is funded by general revenues, requiring it to compete for resources with hundreds of other priorities in the annual appropriations process. Annual appropriations is influenced by the country's fiscal situation at the time, emergencies, changing priorities reflected by a new president or Members of Congress, and individual projects in specific districts and states. These dynamics work against consistency and reduce the chances for sustained progress. Increasing spending one year and reducing it the next is not a sensible public policy. A mechanism needs to be created that would raise

---

[83] The American Energy Innovation Council. *Restoring American Energy Innovation Leadership: Report Card, Challenges and Opportunities.* 2015. P. 6.

[84] Anadon, L. D., M. Bunn, G. Chan, M. Chan, C. Jones, R. Kempener, et al. *Transforming U.S. Energy Innovation.* P. 222.

[85] The American Energy Innovation Council. *A Business Plan.* 2010. P. 20.

[86] The American Energy Innovation Council. *A Business Plan.* 2010. P. 23.

revenue and devote it to funding energy RD&D.[87, 88] This could ensure consistency of funding. Other issues that would need to be addressed in addition to creating the mechanism include how the revenue would be disbursed, who decides what gets funded, and the level of Congressional involvement.[89]

The US Needs to Increase International Collaboration in Energy RD&D

There is a need for the US to increase international collaboration in energy RD&D. Regarding the need to rationalize its international activities, nine US cabinet departments and 10 agencies are involved in the implementation of 175 bilateral arrangements and more than 20 multilateral agreements designed to stimulate energy innovation.[90] In addition to government, the private sector, national laboratories, and other types of entities are engaged in these activities. This needs to be rationalized. The US also needs to increase its collaboration to make better use of the limited resources available for this purpose and to achieve other co-benefits including energy security and to participate in the growing market for low and non-emitting energy technologies.[91, 92] The IEA also identifies specific areas for international collaboration within the context of the gap technologies cited previously.[93]

---

[87] The American Energy Innovation Council. *Restoring American Energy Innovation Leadership.* 2015. P. 11.

[88] There are several mechanisms available that could be utilized to fund increased RD &D. See Nordhaus, R., K. Danish, R. Rosenzweig, P. Runci., G. Stokes, S. Peabody, et al. *Public Sector Funding Mechanisms to Support the implementation of a U.S. Technology Strategy.* Global Energy Technology Strategy Program. GTSP Working Paper 2004–07 (PNNL-14780), 2004.

[89] Nordhaus, R., K. Danish, R. Rosenzweig, et al. *Public Sector Funding Mechanisms to Support the implementation of a U.S. Technology Strategy.*

[90] Anadon, L. D., M. Bunn, G. Chan, M. Chan, C. Jones, R. Kempener, et al. *Transforming U.S. Energy Innovation.* PP. 284–285.

[91] Edmonds, J., T. Wilson, and R. Rosenzweig. 2000. *A Global Energy Technology Strategy Project Addressing Climate Change.* Joint Global Change Research Institute. PP. 50–51.

[92] Anadon, L. D., M. Bunn, G. Chan, M. Chan, C. Jones, R. Kempener, et al. *Transforming U.S. Energy Innovation.* P. 43.

[93] ©OECD/IEA 2011, Clean Energy Progress report—IEA Input to the Clean Energy Ministerial, IEA Publishing. Licence: https://www.iea.org/t&c/. PP. 15–31.

Increased collaboration can take place within existing international organizations or bilateral or multilateral clubs that have been cited.

The US Requires a Long-Term Energy RD&D Strategy

This recommendation focuses on the need for government to make an increased and sustained investment in energy RD&D to develop a portfolio of energy technologies necessary to achieve long-term climate and other important objectives. However, similar to all complex challenges, increased resources are only a partial answer. A long-term strategy to develop and deploy energy technologies in the US must accompany a sustained increase in funding. Some of the issues which need to be considered in developing and implementing a long-term technology strategy and common to all countries include the roles of the sectors included in the innovation process such as government, business, the research community, and academia; the allocation of expenditures among basic research, commercialization, and deployment; the composition of the portfolio; metrics to measure the effectiveness of investment; and the ability to shift resources into priority areas when circumstances dictate. In addition, the government must know when to reallocate resources from technologies that are competitive in the marketplace while increasing support for others such as energy storage, CCS, and new nuclear technolgies that may play a key role in the long-term effort to address climate change.

Policies to Complement RD&D

Consistent and increased investment in energy RD&D and the implementation of a long-term strategy are necessary to improve the performance and to lower the costs of existing energy technologies and to develop new ones to achieve climate policy objectives and other co-benefits. In addition to RD&D, other policies are required to create demand for them and reduce risk. Examples of such policies and which are referred to in the next set of recommendations in a US context include loan guarantees, performance standards, renewable portfolio standards, procurements, and tax policy to name a few.

Recommendation 6: Improve Carbon Intensity and Energy Efficiency in the US

The prior recommendations are applicable in the US, many countries, and potentially at the international level. Because of my expertise and experience, those that follow are limited to the US. They primarily aim to increase the development and deployment of low- and non-emitting energy technologies to reduce CO_2 emissions, as they account for approximately 80% of US GHG emissions.[94] I recognize that this will be an important emphasis of other countries' climate change strategies. However, others are much more qualified to comment on such efforts and to make recommendations in this context.

The context for the recommendations is historic US emissions' performance from 2000 to 2005 where indicated, 2014 data on energy production and consumption, and 2040 projections. In addition to reducing CO_2 emissions, it is important to note that the recommendations would also achieve multiple societal objectives. These include increasing the flexibility of energy policy and national security by expanding the nation's portfolio of technologies, and improving air quality and public health while positioning the US to compete in growing markets for new technologies.

Increase the Development and Deployment of Renewables in the USA

The US power sector remains the largest emitter of GHGs in the US. However, CO_2 emissions have declined more than 10% from 2000 to 2014[95] and 15% from the 2005 levels.[96] They are projected to grow slightly

---

[94] US Environmental Protection Agency. *Inventory of U.S. Greenhouse Gas Emissions and Sinks: 1990–2013.* EPA 430-R-15-004. 2015. ES-9.

[95] US Energy Information Administration. *December 2015 Monthly Energy Review. Carbon Dioxide Emissions From Energy Consumption: Electric Power Sector.* US Department of Energy. DOE/EIA-0035(2015/12), 2015. P. 181.

[96] US Department of State. *United States Climate Action Report 2016, Second Biennial Report of the United States of America, Under the United Nations Framework Convention on Climate Change.* 2016. P. 11.

through 2040, although this does not assume implementation of the CPP.[97] Falling emissions are the result of US electricity generation becoming less carbon-intensive. Since 2000, 93% of new capacity has been gas, wind, solar, or other renewables.[98] Renewables share of power generation has increased from 2007 to 2014 mostly due to strong growth in wind and solar.[99]

The increases in renewables use can be attributed to a combination of federal and state policies. At the federal level, rulemakings designed to achieve reductions of conventional air pollutants and uncertainty regarding climate change policy has made coal less attractive as a long-term fuel. Research and development, tax incentives, and other policies have helped lower the costs of renewables and improved project economics. In addition, 37 states have implemented renewable portfolio standards that require generators to provide a specified amount of their power with renewables.[100] Based on existing policies, renewables are forecast to meet nearly 40% of demand through 2040 and increase their share of generation to 18%.[101]

To ensure continued increases in the development and deployment of renewables, federal investment in RD&D should be increased, tax incentives and regulatory policies should be maintained, and these policies should be updated and redirected when necessary. A few items are worth noting regarding these issues. The allocation of RD&D needs to strike the balance between investing in the key longer-term technologies vital to increasing the use of renewables, such as those that may enable large-scale storage, while reducing support for industries that will be competitive in their own right. The same holds true for tax incentives. And this is challenging as industries attempt to hold on to preferred policies.

Consistency of policy is also important. As mentioned previously, US support for RD&D has been inadequate and inconsistent. A dedicated funding mechanism is required. Similarly, tax policies aimed at boosting

---

[97] US Energy Information Administration, *Annual Energy Outlook 2015 with projections till 2040*. US Dept. of Energy. DOE/EIA—0383(2015), 2015. ES-8.

[98] Zindler, E., M. Di Capua et al. *2015 Factbook: Sustainable Energy in America*. Bloomberg New Energy Finance, The Business Council for Sustainable Energy. 2015. P. 7.

[99] Zindler, E., M. Di Capua et al. *2015 Factbook: Sustainable Energy in America*. P. 19.

[100] US Department of State. *United States Climate Action Report 2016*. P. 28.

[101] US Energy Information Administration, *Annual Energy Outlook 2015 with projections till 2040*. PP. 25–26.

renewables have also been inconsistent. Large-scale solar has benefitted from a federal investment tax credit and it had been scheduled to decline in value in 2016.[102] The Wind Production Tax Credit has expired five times since the end of 2012, each time adversely affecting the deployment of wind. This is a primary reason why 0.5 gigawatts of wind were added in 2013 compared with the nearly five that were added in 2014.[103] Inconsistent polices are an enemy of the continuous improvement in technological performance and certainty that investors require to evaluate and make investments. Most recently, in the context of a budget agreement in the US, clarity was provided regarding important renewable tax incentives for the next several years.[104] This will continue the momentum of increased deployment of renewables in the US. Also, assuming the CPP survives legal scrutiny and political attacks, it too will create an incentive for the addition of more renewables. Analysis that assumes implementation of the CPP concludes that renewables increase to more than 20 % of US generation by 2025.[105] At the state level, RPS programs should be continued and strengthened. Increased levels of coordination between federal and state policies would also benefit renewables.

Last, I would hope that a clean energy standard could be adopted at the federal level setting a goal for clean energy use in the US. In addition to benefitting renewables, a program could be structured to benefit other non-emitting technologies such as nuclear and carbon sequestration. This could increase a program's political viability.

Improve Energy Efficiency

The US economy is becoming more energy efficient. The US GDP per unit of energy consumed improved by 11 % from 2007 to 2014.[106] Good progress has been made in all sectors including electricity,[107]

[102] Zindler, E., M. Di Capua et al. *2015 Factbook: Sustainable Energy in America.* P. 18.

[103] Zindler, E., M. Di Capua et al. *2015 Factbook: Sustainable Energy in America.* P. 9.

[104] See Sections 301 through 304 of the Consolidated Appropriations Act, 2016.

[105] ©OECD/IEA 2015, World Energy Outlook Special Report on Energy and Climate, IEA Publishing. Licence: https://www.iea.org/t&c/ P. 45.

[106] Zindler, E., M. Di Capua et al. *2015 Factbook: Sustainable Energy in America.* P. 7.

[107] Zindler, E., M. Di Capua et al. *2015 Factbook: Sustainable Energy in America.* P. 7.

transportation, industry, and the commercial and residential sectors.[108] This has contributed to falling emissions in these sectors.[109] The Obama Administration has been aggressive in putting policies into place to improve energy efficiency including landmark fuel economy and GHG standards for various types of vehicles, efficiency standards for many energy using products, and building codes, and labeling programs designed to provide consumers with information regarding products energy use.[110, 111]

In addition to federal policies, more than 20 states put programs in place requiring utilities to reduce energy use by a specified amount or percentage each year.[112, 113] Some states have coupled these policies with others that decouple utility rates from sales.[114]

Projections to 2040 indicate continued improvements in energy efficiency. Energy use per 2009 dollar of GDP declines 2 % per year from 2013 to 2040 while carbon intensity declines by 2.3 % per year.[115] Greater improvements could occur as the projections are based on current policies and do not assume implementation of the CPP, which would increase efficiency in the power sector. Because economic growth and low energy prices are assumed to continue during the projection period, additional policies will be required to improve on the projections. These include increased RD&D to improve energy-efficient technologies in the residential, commercial, and industrial sectors, increased standards to stimulate innovation, more stringent building codes, and tax policy. In the transportation sector, a combination of policies is required to build on and

---

[108] US Department of State. *United States' Climate Action Report 2016.* PP. 11–13.

[109] US Department of State. *United States' Climate Action Report 2016.* PP. 11–13.

[110] For a review of energy efficiency policies, see US Department of State, *United States' Climate Action Report 2016. Appendix 3:* U.S. Policies and Measures.

[111] For progress made in the efficiency area see, *Section III of the 2nd Anniversary Progress Report of President Obama's Climate Action Plan.*

[112] US Department of State. *United States' Climate Action Report 2016.* P. 28.

[113] DSIRE NC Clean Energy Technology Center. *Energy Efficiency Resource Standards and Goals. Available from:* http://ncsolarcen-prod.s3.amazonaws.com/wp-content/uploads/2015/03/Energy-Efficiency-Resource-Standards.pdf [Accessed 23 December 2015].

[114] Zindler, E., M. Di Capua et al. *2015 Factbook: Sustainable Energy in America.* P. 101.

[115] US Energy Information Administration, *Annual Energy Outlook 2015 with projections till 2040.* PP. 16–17.

expedite progress that has been achieved. These include more stringent rules increasing fuel efficiency and tightening GHG emissions standards for post-model year 2025 vehicles, increased RD&D to bring down the cost and improve the performance of battery and other long-term technologies necessary to improve the sectors' performance and tax policies. These efforts will need to be coordinated with state policies.

Maintain the Nuclear Option

Nuclear power has continued to play an important role in the US energy system. There are currently 99 nuclear reactors operating, which provided nearly 20 % of US power supply in 2014.[116] These plants accounted for more than 60 % of US emissions free generation in 2014, avoiding nearly 600 million metric tonnes of CO_2 emissions and more than 13 billion metric tonnes from 1995 to 2014.[117] This represents nearly two years of US emissions. Analysis indicates that nuclear power's share of generation is estimated to decline by 2040.[118] About 59 new reactors would need to be built by 2040 for nuclear power to maintain its 20 % share of generation.[119] At the present time, five new reactors are under construction.

Nuclear power is facing a host of economic and policy challenges as well as public concern and skepticism. The economic challenges include low natural gas prices and reactor costs. Policy challenges include an aging fleet, averaging nearly 34 years old.[120] Each plant is initially provided a 40-year operating license and licensees are allowed to apply for a 20-year extension of its license. About 76 plants have already secured

---

[116] US Energy Information Administration, Annual Energy Outlook 2015 with projections till 2040. P. 25.

[117] Nuclear Energy Institute Knowledge Center. *Environment: Emissions Avoided.* Available from: http://www.nei.org/Knowledge-Center/Nuclear-Statistics/Environment-Emissions-Prevented/ Emissions-Avoided-by-the-US-Nuclear-Industry [Accessed 23 December 2015].

[118] US Energy Information Administration. *Annual Energy Outlook 2015 with projections till 2040.* P. 25.

[119] Redmond, E. *Nuclear Power Trends.* [Presentation] to the Global Nexus Initiative. Washington, DC. 22 September 2006.

[120] US Energy Information Administration. *Frequently Asked Questions: How old are US nuclear power plants and when was the last one built?* Available from: https://www.eia.gov/tools/faqs/faq. cfm?id=228&t=21 [Accessed 23 December 2015].

an extension and 22 others either are in the process of applying for a license extension or have stated their intent to do so.[121] The government has also failed to develop a permanent storage site to dispose of spent nuclear fuel, even though billions of dollars were spent attempting to do so. The owners of the plants keep the waste on site causing local concerns. These dynamics impose pressure on the existing fleet and make it increasingly difficult to build new units. In addition to these issues, the public continues to remain concerned with nuclear safety in the aftermath of Fukushima and proliferation risks.

Policies that could contribute to the maintenance of existing nuclear power plants include providing the guidance necessary for a second 20-year renewal of an operating license and requiring the federal government to review the license application for the Yucca Mountain nuclear waste facility. If the potential for new units is going to be maintained, nuclear plants should be eligible to participate in DOE's loan guarantee program for clean energy technologies. Some of the new units under construction in the US received such a guaranty. Another important policy is to increase R&D exploring the potential for smaller, more modular nuclear reactors. This next generation of reactors requires a lower initial investment, produces less waste, and poses less proliferation risk. In addition, as stated previously, nuclear power could benefit if a clean energy standard was adopted at the federal level.

Enable the Continued Use of Coal

Coal use has been in decline in the US. It accounted for more than 50% of power supply in 2000, declining to less than 40% in 2013.[122] Causes for this include low natural gas prices, high up-front capital costs, regulations requiring reductions in conventional and hazardous air pollutants, and uncertain climate change policy. Its long-term prospects are

---

[121] Nuclear Energy Institute Knowledge Center. *US Nuclear License Renewal Filings*. Available from: http://www.nei.org/Knowledge-Center/Nuclear-Statistics/US-Nuclear-Power-Plants/US-Nuclear-License-Renewal-Filings [Accessed 23 December 2015].

[122] US Energy Information Administration. *Annual Energy Outlook 2015 with projections till 2040*. P. 24.

not much better. Analysis estimates that coal declines to approximately 30 % as a fuel for power generation in 2040 if the Clean Power rule is not implemented. Approximately 15 % of capacity is retired and few new plants are built.[123] If the CPP is implemented, coal use declines to less than 25 % by 2025.[124]

Although coal use has declined significantly, it still maintains its share as the largest contributor to US electricity in 2040 according to some analysis.[125] It is also estimated that the US has 261 years of recoverable reserves.[126] The USA should increase its investment in RD&D of CCS and other efficient technologies to maintain coal's continued viability because of its importance to US power supply and contributions to national security. Because of the scale of investment that is required to develop CCS plants, the government could also participate in demonstration projects with industry. Government support for the development of pipeline infrastructure and continued improvement in the certainty of regulatory treatment of CCS could also greatly benefit CCS development.

Recommendation 7: Aggressively Target Non-CO$_2$ GHGs in the US

As discussed in Chap. 6, non-CO$_2$ gases account for 18 % of US emissions. These gases, known as short-lived climate pollutants do not remain in the atmosphere as long as CO$_2$ and actions to reduce them have rapid benefits.

Chapter 6 highlighted some of the Obama Administration's rulemakings targeting these gases in the US and efforts made to coalesce the international community to reduce them. The US should continue to take

---

[123] US Energy Information Administration. *Annual Energy Outlook 2015 with projections till 2040.* PP. 24–26.

[124] ©OECD/IEA 2015, World Energy Outlook Special Report on Energy and Climate, IEA Publishing. Licence: https://www.iea.org/t&c/. P. 45.

[125] US Energy Information Administration. *Annual Energy Outlook 2015 with projections till 2040.* P. 24.

[126] US Energy Information Administration. *Energy Explained: How Much Coal is Left?* Available from: https://www.eia.gov/energyexplained/index.cfm?page=coal_reserves [Accessed 23 December 2015].

action to reduce non-CO_2 gases. Emissions within some of the categories of non-CO_2 gases including methane, HFCs, and NOx emissions are continuing to increase.[127] These represent targets of opportunity to achieve further reductions that could provide important climate benefits.

Conclusion

The need to address the risks created by climate change is urgent. Its impacts are being felt around the globe. It will be extremely challenging to achieve the international community's goal of limiting temperature increases to well below 2 °C. We know that doing so will require enormous reductions in GHG emissions throughout the century within the context of a much larger population and global economy. Dramatic and sustained improvements in carbon and energy intensity are the only way forward.

The world is in this situation because the first generation of climate change policy failed. GHG emissions and concentrations increased significantly from the KP's 1990 base year to the present time. Twenty years were lost negotiating it and implementing policies that did not do the job. I have provided my views of the reasons why the first generation of policy-making failed. There is nothing more to be said regarding the KP's inability to work, given the sliver of emissions it covered when it was agreed to and the inability of its top-down policy framework and the EU ETS to learn from experience, adapt to new information, and address problems in a timely fashion. This is a flaw found in many top-down systems implemented by large institutions, whether in the public or private sector.

Climate change policy 1.0 ended in the US with the defeat of cap-and-trade in the 2009–2010 time frame. And although it is harder to pinpoint when it ended at the international level, it was clear the KP was never up to the task. Officially, the international community decided in 2007 to craft a successor to the KP.

---

[127] US Environmental Protection Agency. *Inventory of U.S. Greenhouse Gas Emissions and Sinks: 1990–2013.* EPA 430-R-15-004. 2015.

In light of prior experience, I have argued that an entirely new approach to GHG emissions mitigation is required; one that is more bottom-up and modest than the initial effort. Small ball as I have previously called it. Why has someone who was so confident in the ability of ambitious, top-down market-based systems to address climate change come to this view? The primary reason is that many of political systems around the world, including in the US, are broken and dysfunctional. This is evidenced by a multitude of persistent problems that remain unattended. Governments have generally proven themselves incapable of solving big problems like climate change. Because of this, citizens frequently have little confidence or trust in their public institutions. I argued earlier that results are the only way to win back the public's trust.

The public's support for climate change policies will increase when the economic and environmental benefits of those policies become evident. Only then will the political conditions exist to undertake the more ambitious actions that will be required to address climate change throughout the twenty-first century. And given the magnitude of the challenge, greater ambition will be necessary as the century marches forward. This is evidenced by analyses of the impacts of the initial [I]NDCs in limiting temperature increases, and the need for GHG emissions to be net zero or negative by 2100.

The new era of climate change policy as described in Chap. 6, is well underway. Both the US strategy to reduce its GHG emissions and the Paris Agreement reached at COP-21 give me reason to be optimistic. With taxes and cap-and-trade off the table in the US, the Obama Administration has been creative in using authority provided by existing laws to reduce GHG emissions from the power and transportation sectors, increase the deployment of renewables, improve efficiency, and reduce non-CO_2 gases. These policies have the potential to continue the trends toward improved carbon and energy intensity in the US.

I believe there is much to like in the Paris Agreement, including its bottom-up approach to mitigation, top-down provisions designed to ensure countries are living up to their commitments, and efforts to stimulate greater ambition. Most importantly, there is the potential that the policy framework embodied in the Paris Agreement can be a building block for consistent, more ambitious actions. One reason for the KPs failure was the uncertainty as to what would occur after the five-year commitment period ended. Business did not know what would come next, if anything, and this served as a disincentive to taking action. In contrast, the processes included in the Paris Agreement, which require Parties to develop successive [I]NDCs and evaluate the collective progress made in achieving the agreements' long-term goals, may provide a signal that governments are serious in addressing climate change and that the issue will be front and center on the international agenda for the foreseeable future. Hopefully, this provides business with the certainty and the incentives that are required to make the long-term investments that are required to decarbonize the economy throughout the twenty-first century.

I am not naïve—and in fact I am frequently labeled as a cynic by friends and colleagues. I recognize that politics may intervene in the US. A new president in the US may attempt to undo the actions of its predecessor, President Obama, and/or place less emphasis on climate change. The centerpiece of the US strategy, the CPP, must withstand legal challenge and ongoing political attacks. The cooperation that characterized the negotiations at COP-21 and resulted in the Paris Agreement may cease. The negotiations may bog down as they had previously. If this occurs, and the agreement is not implemented, its promise will not be realized. Persistent, tough diplomacy will be required to operationalize the concepts that have provided many with optimism that we may be at a turning point.

The potential for problems are limitless. However, for now I remain optimistic. I believe that policy-makers and negotiators have learned from the previous failure, requiring that they throw out the prior playbook and create a new approach. Ultimately, the success of this era of policy-making will be measured by falling emissions, slowing concentrations, and the mitigation of the risks and threats posed by global climate change.

Bibliography

1. Administrations Economic Analysis. (1998). *The Kyoto Protocol and the President's policies to address climate change.*
2. Abbot, K. W. (2011). The transnational regime complex for climate change. *Environment and Planning C: Government and Policy*, Forthcoming. Available at SSRN: http://ssrn.com/abstract=1813198 or http://dx.doi.org/10.2139/ssrn.1813198
3. Akihisa, K., & Kentaro, T. *IGES CDM project database.* Available from: http://pub.iges.or.jp/modules/envirolib/view.php?docid=968
4. Alexeew, J., Bergset, L., Meyer, K., Petersen, J., Schneider, L., & Unger, C. (2010). An analysis of the relationship between the additionality of CDM projects and their contribution to sustainable development. *International Environmental Agreements: Politics Law and Economics, 10*(3), 233–248.
5. American Energy Innovation Council. (2010). *A business plan for America's energy future.* Washington, DC: American Energy Innovation Council.
6. American Energy Innovation Council. (2015). *Restoring American energy innovation leadership: Report card, challenges and opportunities.* Washington, DC: American Energy Innovation Council.
7. Anadon, L. D., Bunn, M., Chan, G., Chan, M., Jones, C., Kempener, R., et al. (2011). *Transforming U.S. energy innovation.* Cambridge, MA: Harvard Kennedy School, BELFER Center for Science and International Affairs.

© The Author(s) 2016
R.H. Rosenzweig, *Global Climate Change Policy and Carbon Markets*, Energy, Climate and the Environment,
DOI 10.1057/978-1-137-56051-3

8. Andresen, S. (2015). International climate negotiations: Top-down, bottom-up or a combination of both? *The International Spectator, 50*(1), 15–30.

9. Asia Pacific Economic Cooperation. (2009). *Singapore declaration: Sustaining growth, connecting the region.*

10. Bartosiewicz, P., & Miley, M. (2013). *The too polite revolution: Why the recent campaign to pass comprehensive climate legislation in the United States failed.* Prepared for the symposium on: The politics of America's fight against global warming, Co-sponsored by the Columbia School of Journalism and the Scholars Support Network, 14 February 2013, Report Commissioned by Lee Wasserman at the Rockefeller Family Fund in conjunction with Nick Lehman, Columbia School of Journalism.

11. Berghhmans, N. (2012). *Energy efficiency, renewable energy and CO_2 allowances in Europe: A need for coordination.* CDC Climate Research, No. 18.

12. Berghmans, N., Sartor, O., & Stephan, N. (2013). *Reforming the EU ETS: Give it some work!* CDC Climate Research, No. 28.

13. Bodansky, D., & Diringer, E. (2014). Alternative models for the 2015 climate change agreement. *FNI Climate Policy Perspectives, 13*, 1–8.

14. Borenstein, S. (2001, March 14). Bush changes pledges on emissions. *Philadelphia Inquirer.*

15. Brown, E. G. Jr. (2015). *Executive order B-30-15.* Available from: https://www.gov.ca.gov/news.php?id=18938

16. Brown, L. M., Hanafi, A., & Petsonk, A. (2012). *The EU emissions trading system: Results and lessons learned.* New York: Environmental Defense Fund.

17. Bureau of Labor Statistics. *The employment situation: December 2008.* [Press Release]. 9 January 2009.

18. Bush, G. W. (2001, March 13). *Letter to members of the Senate on the Kyoto Protocol on climate change.* Online by Gerhard Peters and John T. Woolley, The American Presidency Project. http://www.presidency.ucsb.edu/ws/?pid=45811

19. Byrd, R. C., et al. S. Res. 98, Expressing the sense of the Senate regarding the conditions for the United States becoming a signatory to any international agreement on greenhouse gas emissions under the United Nations Framework Convention on Climate Change. Available at Government Printing Office Web Site: http://www.gpo.gov/fdsys/pkg/BILLS-105sres98ats/pdf/BILLS-105sres98ats.pdf

20. CAIT Climate Data Explorer. (2015). Washington, DC: World Resources Institute. Available online at: http://cait.wri.org

21. California Air Resources Board. (2008). *Climate change scoping plan, A framework for change pursuant to AB 32, The Global Warming Solutions Act.*

22. California Air Resources Board. (2001, July). *Status of plan measures.*

23. California Air Resources Board. (2011, July). *Status of scoping plan recommended measures.*

24. California Air Resources Board. (2014). *First update to the climate change scoping plan, building on the framework pursuant to AB 32, The Global Warming Solutions Act.*

25. Capoor, K., & Ambrosi, P. (2006). *State and trends of the carbon market 2006.* Washington, DC: The World Bank and the International Emissions Trading Association.

26. Capoor, K., & Ambrosi, P. (2007). *State and trends of the carbon market 2007.* Washington, DC: The World Bank and the International Emissions Trading Association.

27. Capoor, K., & Ambrosi, P. (2008). *State and trends of the carbon market 2008.* Washington, DC: The World Bank.

28. Capoor, K., & Ambrosi, P. (2009). *State and trends of the carbon market 2009.* Washington, DC: The World Bank.

29. CDM Insights. *Project activities.* Available from: http://cdm.unfccc.int/Statistics/Public/CDMinsights/index.html

30. Clements, B., Coady, D., Fabrizio, S., et al. (2013). *Energy subsidy reform: Lessons and implications.* Washington, DC: International Monetary Fund. Available from: http://www.eisourcebook.org/cms/March_2013/Energy%20Subsidy%20Reform,%20Lessons%20&%20Implicatio ns.pdf

31. Climate Action Tracker. *2.7°C is not enough—We can get lower.* Available from: http://climateactiontracker.org/assets/publications/briefing_papers/CAT_Temp_Update_COP21.pdf

32. *Climate and clean air coalition to reduce short lived climate pollutants.* Available from: http://www.ccacoalition.org/en

33. Clinton, B., & Gore, A. (1992). *Putting people first: How we all can change America.* New York: Three Rivers Press.

34. Clinton, William J. (1993). Address before a joint session of congress on administration goals. February 17, 1993. Online by Gerhard Peters and John T. Woolley, The American Presidency Project. http://www.presidency.ucsb.edu/ws/?pid=47232

35. Clinton, William J. (1993). Remarks on Earth Day. April 21, 1993. Online by Gerhard Peters and John T. Woolley, The American Presidency Project. http://www.presidency.ucsb.edu/ws/?pid=46460

36. Clinton, W. J., & Gore, A. Jr. (1993). *The climate change action plan.* Washington, DC: Executive Office of the President.

37. Clinton, William J. (1997, October 22). Remarks at the National Geographic Society. Online by Gerhard Peters and John T. Woolley, The American Presidency Project. http://www.presidency.ucsb.edu/ws/?pid=53442

38. Coady, D., Parry, I., Sears, L., & Shang, B. (2015). *IMF working paper: How large are global energy subsidies?* Washington, DC: International Monetary Fund, WP/15/05.

39. Cogen, J., Rosenzweig, R., & Varilek, M. (2003). Overview of emerging markets for greenhouse gas commodities. In *Greenhouse gas market 2003: Emerging but fragmented.* Geneva: International Emissions Trading Association.

40. Cogen, J., Forrister, D., Rosenzweig, R., Vickers, P., & Youngman, R. (2004). Overview of emerging markets for greenhouse gas commodities: Understanding the role of risk in pricing. In R. Dornau (Ed.), *Greenhouse gas market 2004: Ready for takeoff.* Geneva: International Emissions Trading Association.

41. Commission of the European Communities. Climate change—Towards an EU Post Kyoto Strategy. COM (1998). 353 Final, Brussels, 3 June 1998.

42. Commission of the European Communities. Green paper on greenhouse gas emissions trading within the European Union. COM (2000) 87, Brussels, 8 March 2000.

43. Commission Staff Working Document. *Impact assessment accompanying the document, communication from the Commission to the European Parliament, the Council, the European Economic and Social Committee and the Committee of the Regions, A policy framework for climate and energy in the period from 2020 up to 2030.* {COM(2014) 15 final} {SWD(2014) 16 final}.

44. Communication from the Commission to the Council and the Parliament. Preparing for Implementation of the Kyoto Protocol. COM (1999) 230, 19 May 1999.

45. Congressional Budget Office. (2009). *The estimated costs to households from the cap-and-trade provisions of H.R. 2454.* Washington, DC: US Congress.

46. Congressional Budget Office and the Joint Committee on Taxation. (2009, June 26). *Estimated changes in revenues and direct spending under H.R. 2998.* As Amended by the House Committee on rules on June 26, 2009. Washington, DC: US Congress.

47. The Contract with America is available from: http://www.nationalcenter.org/ContractwithAmerica.html

48. Consolidated Appropriations Act of 2016. (2016). *Sections 301 through 304 of Title III, miscellaneous provisions of Division P, tax related provisions.*

49. Cormier, L., & Lowell, R. (2001, September 21). DuPont and Marubeni Execute first UK greenhouse gas emissions allowance trade. [Press Release].

50. de Dominicis, A. (2005, November). *Carbon investment funds: Growing faster* (Caisse des Depots. Research report No. 7). Paris, France.

51. de Perthius, C. (2011). *Carbon markets regulation: The case for a CO_2 Central Bank.* Climate Economic Chair CDC Climate and Paris Dauphine University, No. 10. Paris, France.

52. DG CLIMA. *EU ETS FAQ: Question 20.* Available at: http://ec.europa. eu/clima/policies/ets/faq_en.htm

53. Directive 2009/28/EC of the European Parliament and of the Council of 23 April 2009 on the promotion of the use of energy from renewable sources and amending and subsequently repealing Directives 2001/77/EC and 2003/30/EC. Brussels, European Parliament and Council.

54. Directive 2012/27/EU of the European Parliament and of the Council of 25 October 2012 on energy efficiency, amending Directives 2009/125/ EC and 2010/30/EU and repealing Directives 2004/8/EC and 2006/32/ EC. Brussels, European Parliament and Council.

55. Dr. Pieter Tans, NOAA/ESRL (www.esrl.noaa.gov/gmd/ccgg/trends/) and Dr. Ralph Keeling, Scripps Institution of Oceanography (scrippsco2.ucsd.edu/)

56. DSIRE NC Clean Energy Technology Center. *Energy efficiency resource standards and goals.* Available from: http://ncsolarcen-prod.s3.amazonaws.com/ wp-content/uploads/2015/03/Energy-Efficiency-Resource-Standards.pdf

57. Edison Electric Institute. (1998). *Everyone has a responsibility to protect the environment. Washington, DC.*

58. Edmonds, J. (2015, September 22). *Getting to 2°C: Global goals and realistic responses.* [Presentation] to the Global Nexus Initiative. Washington, DC.

59. Edmonds, J., Dooley, J. J., Malone, E. L., Clarke, L. E., Kim, S. H., Lurz, J. P., et al. (2007). *A global energy technology strategy project addressing climate change: Phase 2 findings from an international public-private sponsored research program.* Richland, WA Pacific Northwest National Laboratory, PNNL-SA-5171.

60. Edmonds, J., Wilson, T., & Rosenzweig, R. (2000). *A global energy technology strategy project addressing climate change: An initial report from an international public-private collaboration.* College Park, MD: Joint Global Change Research Institute.

61. Edward, G., & Winum, L. (2002, May 7). Danish electricity supplier Elsam conduct first ever transboundary swap in greenhouse gas compliance instruments. [Press Release]. At the time of this writing, there is no record of this trade on the internet.

62. Eichhammer, W., Boedde, U., Gagelmann, F., Jochem, E., Kling, N., Schleich, J., et al. (2001). *Greenhouse gas reductions in Germany and the UK—Coincidence or policy induced.* Berlin: Federal Environmental Agency, Research Report 201 41 133.

63. Ellerman, D., Marcantonini, C., & Zaklan, A. (2014). *The EU ETS: Eight years and counting.* Robert Schulman Centre for Advanced Studies, European University Institute, RSCAS2014/04.

64. Ellerman, D., Convery, F., & de Perthius, C. (2010). *Pricing carbon: The European Union emissions trading scheme.* New York/Cambridge: Cambridge University Press.

65. Engel, K. (2006). State and local climate change initiatives: What is motivating state and local governments to address a global problem and what does this say about federalism and environmental law. *The Urban Lawyer, 38*(4), 1–17.

66. Energy and Commerce Committee Democrats. Archive of Documents related to HR 2454 the 'American Clean Energy and Security Act' is available at https://wayback.archive-it.org/4949/20141223163051/http://democrats.energycommerce.house.gov/index.php?q=bill/hr-2454-the-american-clean-energy-and-security-act

    a. Floor Action on H.R. 2454. The American Clean Energy and Security Act of 2009. 26 June 2009.

    b. Text of H.R. 2454. The American Clean Energy and Security Act, as passed by the U.S. House of Representatives. 26 June 2009.

    c. Section-by-Section Summary of H.R 2454. The American Clean Energy and Security Act, as passed by the U.S. House of Representatives. 24 July 2009.

    d. Fact Sheet. The American Clean Energy and Security Act of 2009: By the Numbers, 17 September 2009.

    e. Summary of EIA's Economic Analysis of H.R. 2454. The American Clean Energy and Security Act of 2009, 4 August 2009.

    f. Summary of EPA's Economic Analysis of H.R. 2454. The American Clean Energy and Security Act of 2009, 23 June 2009.

    g. CBO: Waxman-Markey costs about a postage stamp a day, Saves Low-Income Families Money.

    h. Full Committee Markup of H.R 2454. The American Clean Energy and Security Act of 2009. 21–24 May 2009.

    i. Chairmen Waxman and Markey Introduce "The American Clean Energy and Security Act," 15 May 2009.

    j. Proposed Emission Allowance Allocation in American Clean Energy and Security Act. 15 May 2009.

    k. Hearing on H.R. 2454, "The American Clean Energy and Security Act of 2009", 24 April 2009.

    l. Chairmen Waxman, Markey Release Discussion Draft of New Clean Energy Legislation, 31 March 2009.

    m. Hearing on "The U.S. Climate Action Partnership," 15 January 2009.

67. Environment, Energy and Natural Resource Options book. This was a document prepared for the Clinton Administration. 1992.

68. Erlander, D. (1994). The BTU tax experience: 'What happened and why it happened. *Pace Environmental Law Review, 12*(1), 173–184.

69. European Commission, *Revised emissions trading system will help the EU achieve its climate goals*. Available from: http://ec.europa.eu/clima/news/articles/news_2015071501_en.htm

70. European Commission. *Revision for Phase IV (2020–2030)*. Available from: http://ec.europa.eu/clima/policies/ets/revision/index_en.htm

71. European Commission. *2030 energy strategy*. Available from: https://ec.europa.eu/energy/en/topics/energy-strategy/2030-energy-strategy

72. European Commission. COMMISSION REGULATION (EU) NO 176/2014 of 25 February 2014 amending Regulation (EU) 1031/2010 in particular to determine the volumes of greenhouse gas emissions allowances to be auctioned in 2013–2020. Available from: http://eur-lex.europa.eu/legal-content/EN/TXT/?uri=uriserv:OJ.L_.2014.056.01.0011.01.ENG

73. European Commission. Communication from the Commission to the European Parliament, the Council, the European Economic and Social Committee and the Committee of the regions. A policy framework for climate and energy in the period from 2020 to 2030. COM (2014) 15 final.

74. European Commission. (2011, January 21). *Emissions trading: Commission welcomes vote to ban certain industrial gas credits*. [Press Release].

75. European Commission. *EU ETS 2005–2012: Evolution of the European carbon market*. Available from: http://ec.europa.eu/clima/policies/ets/pre2013/index_en.htm

76. European Commission. (2010, November 25). *Questions & answers on emissions trading: Use restrictions for certain industrial gas credits as of 2013*. [Press Release].

77. European Commission. Report from the Commission to the European Parliament and the Council. *The state of the European carbon market in 2012*. European Commission. COM(2012) 652 final.

78. European Commission. (2013). The EU emissions trading scheme. *European Commission*. doi:10.2834/55480

79. European Commission. (2013). *The EU Emissions Trading System (EU ETS)*. European Union. Available from: http://ec.europa.eu/clima/publications/docs/factsheet_ets_en.pdf

80. European Commission. *The European Union Emissions Trading System (EU ETS). Questions and answers on the revised EU emissions trading system*. Available from: http://ec.europa.eu/clima/policies/ets/faq_en.htm

81. European Environment Agency. Questions and answers on… key facts about Kyoto targets, June 2010.

82. European Environment Agency. *EU Emissions Trading System (ETS) data viewer*. Available from: http://www.eea.europa.eu/data-and-maps/data/data-viewers/emissions-trading-viewer

83. European Environment Agency. (2014). *Trends and projections 2014, tracking progress towards Europe's climate and energy targets for 2020* (EEA Report No 6/2014).

84. European Parliament Council. (2003). *Directive 2003/87/EC of the European Parliament and of the Council of 13 October 2003 establishing a scheme for greenhouse gas emissions trading within the Community and amending Council Directive 96/61/EC.* Brussels: European Parliament and Council

85. European Parliament Council. (2004). *Directive 2004/101/EC of the European Parliament and of the Council of 27 October 2004 amending Directive 2003/87/EC establishing a scheme for greenhouse gas emission allowance trading within the community in respect, of the Kyoto Protocol's project mechanisms.* Brussels: European Parliament and Council.

86. European Parliament Council. (2009). *Directive 2009/29/EC of The European Parliament and of the Council of 23 April 2009 amending Directive 2003/87/EC so as to improve and extend the greenhouse gas emission allowance trading scheme of the Community.* Brussels: European Parliament and Council.

87. Executive Office of the President. (2015). President Obamas climate action plan, 2nd anniversary progress report. *The White House.*

88. Executive Office of the President. (2013). The President's climate action plan. *The White House.*

89. Falkner, R., Hannes, S., & Vogler, J. (2010). International climate policy after Copenhagen: Towards a 'building blocks approach'. *Global Policy, 1*(3), 252–262.

90. Federal Republic of Brazil. *Brazil's intended nationally determined contribution.* Available from: http://www4.unfccc.int/submissions/INDC/Published%20Documents/Brazil/1/BRAZIL%20iNDC%20english%20FINAL.pdf

91. Fialka, J. (2005, February 28). Natsource forms investment pool to meet greenhouse-gas credits. *Wall Street Journal.*

92. Final Vote for Roll Call 199. *Omnibus Budget Reconciliation Act of 1993.* Available at the Clerk of the House Web Site: http://clerk.house.gov/evs/1993/roll199.xml

93. Final Vote for Roll Call 406. *Omnibus Budget Reconciliation Act of 1993.* Available at the Clerk of the House Web Site: http://clerk.house.gov/evs/1993/roll406.xml

94. Friedrich, J., Ge, M., & Damassa, T. *Infographic: What do your country's emissions look like?* Available from: http://www.wri.org/blog/2015/06/infographic-what-do-your-countrys-emissions-look

95. G-20. (2009). *Leaders Statement.* The Pittsburgh Summit.

96. Garland, C. (1988). Acid rain over the United States and Canada: The D.C. circuit fails to provide shelter under section 115 of the clean air act while state action provides a temporary umbrella. *Boston College Environmental Affairs Law Review, 16*(1), 1–37.

97. Gillenwater, M., & Seres, S. (2011). *The clean development mechanism: A review of the first international offset program.* Arlington, VA: Pew Center on Global Climate Change.

98. Gloaguen, O., & Alberola, E. (2013). *Assessing the factors behind CO_2 emissions changes over Phase 1 and 2 of the EU ETS: An econometric analysis.* Paris: CDC Climate Research, Working paper No. 2013–2015.

99. Green, J. F., Sterner, T., & Wagner, G. (2014). A balance of bottom-up and top-down in linking climate policies. *Nature Climate Change, 4,* 1064–1067.

100. Green, F. J. (2014, December 1). A realistic approach to linking carbon markets. *Washington Post.*

101. Goodell, J. (2015, October 8). Obama takes on climate change. *Rolling Stone Magazine.* Issues 1245.

102. Gore, A. (1992). *Earth in the balance: Ecology and the human spirit.* Emmaus: Rodale.

103. Gorman, S. *Gore's "inconvenient truth" wins documentary Oscar.* Available from: http://www.reuters.com/article/2007/02/26/us-oscars-gore-idUSN 2522150720070226

104. govtrack.us. *Vote on HR2454.* Available from: https://www.govtrack.us/ congress/votes/111-2009/h477

105. govtrack.us. *Vote on S. 1630.* Available from: https://www.govtrack.us/ congress/votes/101-1990/h525

106. Hagel, C., Craig, L., Helms, J., & Roberts, P. (2001, March 6). Letter to President Bush seeking clarification on the Administrations climate change policy.

107. Harris, P. G. (1999). Common but differentiated responsibility: The Kyoto Protocol and US policy. *New York University Environmental Law Journal, 27.*

108. Heinzerling, L. (2007). Massachusetts v. EPA. *The Journal of Environmental Law and Litigation, 22,* 301–311.

109. Barker, M., & Hadi, A. (2010). Payroll employment in 2009: Jobs losses continue. *Monthly Labor Review, 133,* 23–33.

110. Henkemens, M. (2004, June 25). *Dutch lessons as GHG buyer.* [Lecture] New York.

111. Henry J. Kaiser Family Foundation. (2012). *Health care costs: A primer. Key information on health care costs and their impacts.* Number, 7670-03, P. 1.

112. Hinze, J. (2015, March 10). *Global nuclear new build and technology expansion.* [Presentation] to the Nuclear Energy Institute and Partnership for Global Security. Washington, DC.

113. Hovi, J., & Skodvin, T. (2008). Which was to U.S. climate cooperation? Issue Linkage versus a U.S. based agreement. *Review of Policy Research, 25*(2), 129–148.

114. Hovi, J., Sprinz, D., & Bang, G. (2010). Why the United States did not become a party to the Kyoto Protocol: German, Norwegian, and US Perspectives. *European Journal of International Relations, 18*(1), 129–150.

115. IEA. (2011). *Clean Energy Progress report—IEA Input to the Clean Energy Ministerial.* Paris: OECD/IEA.

116. IEA. (2013). *Managing interactions between carbon pricing and existing energy policies.* Paris: OECD/IEA.

117. IEA. (2014). *Energy climate change & environment.* Paris: OECD/IEA.

118. IEA. (2014). *World energy investment outlook special report.* Paris: OECD/IEA.

119. IEA. (2015). *World energy outlook special report on energy and climate.* Paris: OECD/IEA.

120. IEA. (2015). *World energy outlook 2015.* Paris: OECD/IEA.

121. IEA. (2014). *Key world energy statistics.* Paris: OECD/IEA.

122. IEA. (2015). *IEA/IRENA Global renewable energy policies and measures database.* Paris: OECD/IEA.

123. IEA. (2014). CO_2 *Emissions from fuel combustion 2014.* Paris: OECD/IEA.

124. IEA. (2015). CO_2 *Emissions from fuel combustion 2015.* Paris: OECD/IEA.

125. India. *India's intended nationally determined contribution.* Available from: http://www4.unfccc.int/submissions/INDC/Published%20Documents/India/1/INDIA%20INDC%20TO%20UNFCCC.pdf

126. International Atomic Energy Agency. (2015). *Nuclear power reactors in the world, 2015 Edition.* Vienna: International Atomic Energy Agency.

127. International Energy Agency, Organization for Economic Co-operation, Organization for Petroleum Exporting Countries, & World Bank. (2011). *Joint report on fossil fuel and other energy subsidies: An update on the G-20 Pittsburgh and Toronto commitment.*

128. International Emissions Trading Association. (2015, July). *The market stability reserve: Where are we with reform of the EU ETS.*

129. The International Emissions Trading Association, CDC Climate Research, & Environmental Defense Fund. *The world's carbon markets: A case study*

guide to emissions trading. Available from: https://ieta.memberclicks.net/ worldscarbonmarkets; Kelly, C., Helme, N. (2000). *Ensuring CDM project compatibility with sustainable development goals.* Working paper. Washington, DC: Center for Clean Air Policy.

130. IPCC. (2007). *Climate change 2007: Synthesis report. Contribution of Working Groups I, II and III to the fourth assessment report of the Intergovernmental Panel on Climate Change* (p. 5). Core Writing Team, R. K. Pachauri & A. Reisinger (Eds.). Geneva: IPCC.

131. IPCC. (2007). *Climate change 2007: The physical science basis. Contribution of Working Group I to the fourth assessment report of the Intergovernmental Panel on Climate Change.* S. Solomon, D. Qin, M. Manning, Z. Chen, M. Marquis, K. B. Averyt, M. Tignor, & H. L. Miller (Eds.). Cambridge/ New York: Cambridge University Press, 996 pp.

132. IPCC—Synthesis Report Summary for Policymakers. (2014). *Climate change 2014: Synthesis report. Contribution of Working Groups I, II and III to the fifth assessment report of the Intergovernmental Panel on Climate Change.* Core Writing Team, R. K. Pachauri, & L. A. Meyer (Eds.). Geneva: IPCC, 151 pp.

133. IPCC. (2014). Summary for policymakers. In O. Edenhofer, R. Pichs-Madruga, Y. Sokona, E. Farahani, S. Kadner, K. Seyboth, A. Adler, I. Baum, S. Brunner, P. Eickemeier, B. Kriemann, J. Savolainen, S. Schlömer, C. von Stechow, T. Zwickel, & J. C. Minx (Eds.), *Climate change 2014: Mitigation of climate change. Contribution of Working Group III to the fifth assessment report of the Intergovernmental Panel on Climate Change* (p. 10). Cambridge/New York: Cambridge University Press.

134. Kirkman, G., Seres, S, Haites, E., & Spalding-Fecher, R. (2012). *Benefits of the clean development mechanism 2012.* United Nations Framework Convention on Climate Change.

135. Kossoy, A., Peszko, G., Oppermann, K., Prytz, N., Klein, N., Blok, K., et al. (2015, September). *State and trends of carbon pricing 2015.* Washington, DC: World Bank.

136. Kossoy, A., & Ambrosi, P. (2010). *State and trends of the carbon market 2010.* Washington, DC: The World Bank.

137. Kossoy, A., & Guignon, P. (2012). *State and trends of the carbon market 2012.* Washington, DC: The World Bank.

138. Laing, T., Sato, M., Grubb, M., & Comberti, C. (2014). The effects and side-effects of the EU emissions trading scheme. *WIREs Clim Change 2014, 5,* 509–519. doi:10.1002/wcc.283

139. Layzer, J. (2011). Cold front: How the recession stalled Obama's clean energy agenda. In T. Skocpol & L. Jacobs (Eds.), *Reaching for a new deal: Ambitious governance, economic meltdown, and polarized politics in Obama's first two years* (pp. 321–385). New York: Russell Sage Foundation.

140. Leal-Arcas, R. (2012). Top-down and bottom-up approaches in climate change and international trade. X annual conference of the Euro-Latin Study Network on Integration and Trade (ELSNIT) trade and climate change, 19–20 October 2012, Milan.

141. Lecocq, F., & Capoor, K. (2003). *State and trends of the carbon market 2003*. Washington, DC: PCF plus Research, World Bank.

142. Lecocq, F. (2004). *State and trends of the carbon market 2004*. Washington, DC: World Bank.

143. Lecocq, F., & Capoor, K. (2005). *State and trends of the carbon market 2005*. Washington, DC: International Emissions Trading Association and the World Bank.

144. LeCocq, F. (2003). Pioneering transactions, catalyzing markets, and building capacity: The prototype carbon fund contributions to climate policies. *American Journal of Agricultural Economics, 85*(3), 703–707.

145. Lizza, R. (2010, October 11). As the world burns. *The New Yorker Magazine.*

146. Makhijani, S. (2014). *Cashing in on all of the above: U.S. fossil fuel production under Obama*. Washington, DC: Oil Change International.

147. Marcantonini, C., & Ellerman, D. (2013). *The cost of abating CO_2 emissions by renewable energy incentives in Germany*. MIT Center for Energy and Environmental Policy Research. CEEPR WP 2013-005.

148. MacCracken, C. N., Edmonds, J., Kim, S., & Sands, R. (1999). The economics of the Kyoto Protocol. *The cost of the Kyoto Protocol: A multi-model evaluation, A Special Issue of the Energy Journal, 40,* 25–71.

149. McCain, J., & Lieberman, J. The Climate Stewardship Act of 2003.

150. McConnell, M. *Senate Majority Leader Mitch McConnell's Letter to Nation's Governors.* Available from: http://www.mcconnell.senate.gov/public/index.cfm?p=newsletters&ContentRecord_id=d57eba06-0718-4a22-8f59-1e610793a2a3&ContentType_id=9b9b3f28-5479-468a-a86b-10c747f4ead7&Group_id=2085dee5-c311-4812-8bea-2dad42782cd4. Accessed 19 Mar 2015.

151. Mendelson, J., & Kimbrell, A. Petition for rulemaking and collateral relief seeking the regulation of greenhouse gas emissions from new motor vehicles under? 202 of the Clean Air Act. International Center for Technical Assessment. Available from: http://www.ciel.org/Publications/greenhouse_petition_EPA.pdf

152. Murray, B., & Ross, M. (2007). *America's Climate Security Act: A preliminary assessment of potential economic impacts*. Nicholas Institute for Environmental Policy Solutions, Duke University & RTI International, Report number: NI PB 07-04.

153. Natsource Advisory and Research. (2005). *Introduction to Natsource and markets for emissions and renewable energy*.

154. Newell, R. G., Pizer, W., & Raimi, D. (2013, Winter). Carbon markets 15 years after Kyoto: Lessons learned, new challenges. *Journal of Economic Perspectives, 27*(1), 123–146.

155. Nobelprize.org—The Nobel Peace Prize for 2007 to the Intergovernmental Panel on Climate Change (IPCC) and Albert Arnold (Al) Gore Jr. – Press Release. *Nobelprize.org*. Nobel Media AB 2014. Available from: http://www.nobelprize.org/nobel_prizes/peace/laureates/2007/press.html. Accessed 17 Dec 2015.

156. Nordhaus, R., Danish, K., Rosenzweig, R., Runci, P., Stokes, G., Peabody, S., et al. *Public sector funding mechanisms to support the implementation of a U.S. technology strategy*. Working paper 2004–07. (PNNL-14780).

157. Nuclear Energy Institute Knowledge Center. *Environment: Emissions avoided*. Available from: http://www.nei.org/Knowledge-Center/Nuclear-Statistics/Environment-Emissions-Prevented/Emissions-Avoided-by-the-US-Nuclear-Industry

158. Nuclear Energy Institute Knowledge Center. *US Nuclear license renewal filings*. Available from: http://www.nei.org/Knowledge-Center/Nuclear-Statistics/US-Nuclear-Power-Plants/US-Nuclear-License-Renewal-Filings

159. Nuclear Energy Institute Knowledge Center, Environment: Emissions Prevented. http://www.nei.org/Knowledge-Center/Nuclear-Statistics/Environment-Emissions-Prevented

160. Nunez, F., & Pavley, F. *California Global Warming Solutions Act of 2006*. Available from: http://www.leginfo.ca.gov/pub/05-06/bill/asm/ab_0001-0050/ab_32_bill_20060831_enrolled.html

161. Obama08. *Blueprint for change*. Available from: https://ia801003.us.archive.org/19/items/346512-obamablueprintforchange/346512-obamablueprintforchange.pdf

162. Obama, B. (2009, December 18). *Remarks by the President at the morning plenary session of the United Nations Climate Change Conference*. Copenhagen.

163. Plummer, B. (2008, September–October). A new leaf. *Audubon Magazine, 110*(5).

164. OECD. (2015). *OECD companion to the inventory of support measures for fossil fuels 2015*. Paris: OECD Publishing. http://dx.doi.org/10.1787/9789264239616-en

165. Office of Domestic and International Energy Policy. (1993). *Briefing on energy taxes*. Washington, DC: US Department of Energy.

166. Office of the Press Secretary, The White House. (2002, February). *Fact sheet: President Bush announces clear skies & global climate change initiatives*. Washington, DC, Executive Office of the President.

167. Olsen, K. H., & Fenhann, J. (2008). Sustainable development benefits of CDM projects: A new methodology for sustainability assessment based on text analysis of the project design documents submitted for validation. *Energy Policy, 36*, 2819–2830.

168. Olsen, K. H. (2012, December 4–6). *CDM sustainable development co-benefit indicators*. [Presentation] UNEP.

169. Olsen, K. H. (2005). *The clean development mechanisms contribution to sustainable development: A review of the literature*. UNEP Riso Centre: Energy Climate and Sustainable Development, Riso National Laboratory. 2005.

170. On the Conference Report to H. R. 2264. *Omnibus Budget Reconciliation Act of 1993*. Available at: United States Senate Web Site: http://www.senate.gov/legislative/LIS/roll_call_lists/roll_call_vote_cfm.cfm?congress=103&session=1&vote=00247. Accessed 6 Aug 1993.

171. PBL Netherlands Environmental Assessment Agency. *PBL climate pledge INDC tool*. Available from: http://infographics.pbl.nl/indc/

172. Pew Center on Global Climate Change (Currently Center for Climate and Energy Solutions). (2009). *Summary of H.R. 2454: American Clean Energy and Security Act*. Available from: http://www.c2es.org/docUploads/Waxman-Markey%20summary_FINAL_7.31.pdf

173. Pew Center on Global Climate Change (Currently Center for Climate and Energy Solutions). *Summary of the Dingell Boucher Discussion Draft*. Available from: http://www.c2es.org/docUploads/Dingell-Boucher Summary.pdf

174. Platts McGraw Hill Financial. *EU Council adopts CO_2 market reserve to start January 2019*. Available from: http://www.platts.com/latest-news/electric-power/london/eu-council-adopts-co2-market-reserve-to-start-26213345

175. Pollard, Y., & Winum, L. (2001, December 6). *U.S. utility and Danish electricity supplier conduct first trade in Danish greenhouse gas allowances*. [Press Release]. At the time of this writing, there is no record of this trade on the internet.

176. Pooley, E. (2010). *The climate war: True believers, power brokers, and the fight to save the earth.* New York: Hyperion.

177. Prins, S., Galiana, I., Green, C., Grundmann, R., Hulme, M., Korhola, A., et al. (2010). *The Hartwell paper: A new direction for climate policy after the crash of 2009.* London: Institute For Science, Innovation and Society, University of Oxford, LSE Mackinder Programme for the Study of Long Wave Events, The London School of Economics and Political Science.

178. Prins, G., & Rayner, S. (2007). Time to ditch Kyoto. *Nature, 449,* 973–975.

179. Redmond, L., & Convery, F. (2015). The global carbon market-mechanism landscape: pre and post 2020 perspectives. *Climate Policy, 15*(5), 647–669.

180. Redmond, E. (2015). *Nuclear power trends.* [Presentation] at the Global Nexus Initiative Workshop on the Role of nuclear power in a carbon constrained world, Washington, DC, 22 September 2015.

181. Regional Greenhouse Gas Initiative. *Auction 14 results.* Available from: https://www.rggi.org/market/co2_auctions/results/auctions-1-28

182. Regional Greenhouse Gas Initiative. *Auction 3 results.* Available from: https://www.rggi.org/market/co2_auctions/results/auctions-1-28

183. Regional Greenhouse Gas Initiative. *First control period interim adjustment for banked allowances announcement.* Available from: https://www.rggi.org/docs/FCPIABA.pdf

184. Regional Greenhouse Gas Initiative. *Memorandum of understanding.* Available from: http://rggi.org/docs/mou_final_12_20_05.pdf

185. Regional Greenhouse Gas Initiative. *Model rule.* Available from: https://www.rggi.org/docs/model_rule_8_15_06.pdf

186. Regional Greenhouse Gas Initiative. *Overview of RGGI CO$_2$ budget trading program.* Available from: http://www.rggi.org/docs/program_summary_10_07.pdf

187. Regional Greenhouse Gas Initiative. *Second control period interim adjustment for banked allowances announcement.* Available from: https://www.rggi.org/docs/SCPIABA.pdf

188. Regional Greenhouse Gas Initiative. *Summary of RGGI model rule changes.* Available from: https://www.rggi.org/docs/ProgramReview/_FinalProgramReviewMaterials/Model_Rule_Summary.pdf

189. Rosenzweig, R., & Forrister, D. (2001, August 6). *Natsource compiles first comprehensive analysis of the greenhouse gas trading market.* [Press Release].

190. Rosenzweig, R., Varilek, M., Feldman, B., Kuppalli, R., & Janssen, J. (2002). *The emerging international greenhouse gas market.* Arlington: Formerly Pew Center on Global Climate Change and Currently the Center for Climate and Energy Solutions.

191. Rosenzweig, R. (2005, March 3). *Natsource announces launch of the greenhouse gas credit aggregation pool.* [Press Release].

192. Rosenzweig, R. (2005, October 19). *Natsource closes greenhouse gas credit aggregation pool.* [Press Release].

193. Rosenzweig, R., & Youngman, R. (2005). Looking forward from 2005: More surprises to come? In R. Dornau (Ed.), *Greenhouse gas market 2005: The rubber hits the road.* Geneva: International Emissions Trading Association.

194. Rosenzweig, R. (2006, August 29). *Natsource announces participation in the largest greenhouse gas transaction on record.* [Press Release].

195. Rosenzweig, R., Arino, A., & Youngman, R. (2006). Price volatility in EU ETS and other environmental markets. In *Greenhouse gas market 2006: Financing response to climate change.* Geneva: International Emissions Trading Association.

196. Rosenzweig, R., Nelson, E., & Youngman, R. (2007). Provisions impacting costs in U.S. GHG trading proposals. In *Greenhouse gas market 2007: Building upon a solid foundation: The emergence of a global emissions trading system.* Geneva: International Emissions Trading Association.

197. Rosenzweig, R. (2008, March 6). *Natsource recognized as world's largest purchaser of carbon credits by leading investor research firm.* [Press Release].

198. Rosenzweig, R., Youngman, R., & Diamant, A. (2013). *Exploring the interaction between California's greenhouse gas emissions cap-and-trade program and complementary emissions reduction policies.* Electric Power Research Institute. Report number: 3002000298.

199. Royden, A. (2010). U.S. Climate change policy under President Clinton: A look back. *Golden Gate University Law Review, 32*(4), 415–478.

200. Sabel, Charles F., & Victor, D. (2015). Governing global problems under uncertainty: Making bottom-up climate policy work. *Climatic Change,* forthcoming. Earlier version available online at: http://www2.law.columbia.edu/sabel/papers/Sabel%20and%20Victor%20Climatic%20Change%20MAY%2027.pdf

201. Schiermeier, Q. (2012). The Kyoto protocol: Hot air. *Nature, 491*(7426), 656–658.

202. Shishlov, I., & Bellassen, V. (2012). *10 Lessons from 10 years of the CDM.* CDC Climate Research, No. 37.

203. Simendinger, A. (2009, June 5). Will key chairman power up for energy bill? *National Journal.*

204. Sheppard, K. *Conservative activists wage war on Republicans who supported climate bill.* Available from: http://grist.org/article/2009-07-02-cap-and-traitors/

205. Skocpol, T. Naming the problem: What it will take to counter extremism and engage Americans in the fight against global warming: Prepared for the symposium on: *The politics of America's fight against global warming, co-sponsored by the Columbia School of Journalism and the scholars support network, 14 February 2013*. Report Commissioned by the Rockefeller Family Fund in conjunction with Nick Lehman, Columbia School of Journalism.

206. Stephan, N., Bellassen, V., & Alberola, A. (2013). *Use of industrial credits By European industrial installations: From an efficient market to a burst bubble*. CDC Climate Research, No. 43.

207. Sutter, C., & Perreno, J. C. (2007). Does the current clean development mechanism (CDM) deliver its sustainable development claim? An analysis of officially registered CDM projects. *Climatic Change, 84*(1), 75–90.

208. Turner, G. (2012, October 15). *Carbon market update*. [Presentation] to the 12th Annual IEA-IETA-EPRI annual workshop on greenhouse gas emissions trading. Paris..

209. Congressional Record 155 (98). (2009, June 26). *Floor debate on the American clean energy and security act*. H 7619–7687, Available from: https://www.congress.gov/crec/2009/06/26/CREC-2009-06-26-house.pdf PP. H. 7619–7687.

210. United Nations Environmental Programme. (1987). *The Montreal protocol on substances that deplete the ozone layer*. Nairobi: United Nations.

211. United Nations, Framework Convention on Climate Change. (1992). *United Nations Framework Convention on Climate Change*. Bonn: United Nations.

    a. 1995. 1st Session. *Report of the conference of the parties on its first session, held at Berlin from 28 March to 7 April 1995, The Berlin Mandate*. Decision 1/CP.1. United Nations.

    b. 1995. *Report of the conference of the parties on its first session, held at Berlin from 28 March to 7 April 1995, Activities implemented jointly under the pilot phase*. Decision 5/CP.1. United Nations.

    c. 1996. *Report of the conference of the parties on its second session, held at Geneva from 8 to 19 July 1996, The Geneva Ministerial Declaration*. United Nations.

    d. 1998. *Kyoto Protocol to the United Nations framework convention on climate change*. United Nations.

    e. 2002. *Report of the conference of the parties on its seventh session, held at Marrakesh from 29 October to 10 November 2001, Marrakesh Accords*. Draft Decision -/CMP.1. United Nations.

    f. 2006. *Report of the conference of the parties serving as the meeting of the parties to the Kyoto Protocol on its first session, held at Montreal from 28*

November to 10 December 2005. Further guidance related to the clean development mechanism. Decision 7/CMP.1. United Nations.

g. 2007. *Report of the conference of the parties serving as the meeting of the parties to the Kyoto Protocol on its second session, held at Nairobi from 6 to 17 November 2006. Further guidance relating to the clean development mechanism.* Decision 1/CMP.2. United Nations.

h. 2008. *Report of the conference of the parties on its thirteenth session, held in Bali from 3 to 15 December 2007, Bali action plan.* Decision 1/CP.13. United Nations.

i. 2010. *Report of the conference of the parties on its fifteenth session, held in Copenhagen from 7 to 19 December 2009, The Copenhagen Accord.* Decision 2/CP. 15. United Nations.

j. 2012. *Report of the conference of the parties on its seventeenth session, held in Durban from 28 November to 11 December 2011, Establishment of an Ad Hoc working group on the Durban platform for enhanced action.* Decision 1/CP.17. United Nations.

k. 2014. *Report of the conference of the parties on its nineteenth session, held in Warsaw from 11 to 23 November 2013, further advancing the Durban platform.* Decision 1/CP.19. United Nations.

l. 2015. *Report of the conference of the parties on its twentieth session, held in Lima from 1 December to 14 December 2014, Lima call for climate action.* Decision 1/CP.20. United Nations.

m. 2015 *The Paris agreement to the United Nations Framework Convention on Climate Change.* United Nations.

212. United Nations Framework Convention on Climate Change. *GHG emissions profiles for Annex I parties and major groups.* This web site includes data sets for each Annex I party and 3 summary reports. Available from: http://unfccc.int/ghg_data/ghg_data_unfccc/ghg_profiles/items/4625.php

213. United Nations Framework Convention on Climate Change. (2016). *Synthesis report on the aggregate effect of the intended nationally determined contributions.* United Nations. FCCC/CP/2015/7.

214. United Nations Framework Convention on Climate Change. (2016). Submitted INDCs.

215. UNEP DTU Partnership. *CDM/JI pipeline analysis.* Available from: http://www.cdmpipeline.org/

216. United Nations Framework Convention on Climate Change. *Project activities.* Available from: http://cdm.unfccc.int/Statistics/Public/CDM insights/index.html

217. United Nations Framework Convention on Climate Change. *Voluntary tool for describing sustainable development co-benefits (SDC) of CDM project*

activities or programmes of activities (SD Tool). Available at: http://cdm.unfccc.int/Reference/tools/index.html

218. United Nations, Department of Economic and Social Affairs, Population Division. (2015). *World population prospects: The 2015 revision, key findings and advance tables.* Working paper no. ESA/P/WP.241.

219. United Nations. (1997). *Glossary of environmental statistics.* Report number: ST/ESA/STAT/SER.F/67.

220. UNEP DTU Partnership. *Content of CDM/JI pipeline. Approved CDM methodologies.* Available from: http://www.cdmpipeline.org/cdm-methodologies.htm

221. UNFCCC Secretariat. (2007). *Report of the 32nd meeting of the executive board of the clean development mechanism.* UNFCCC. CDM-EB-32.

222. UFCCC Secretariat. (2007). *Report of the 33rd meeting of the executive board of the clean development mechanism.* UNFCCC. CDM-EB-33.

223. UFCCC Secretariat. (2008). *Report of the 41st meeting of the executive board of the clean development mechanism.* UNFCCC. CDM-EB-41.

224. UNEP Green Economy Policy Brief. *Fossil fuel subsidies.* Available from: http://www.unep.org/greeneconomy/Portals/88/documents/GE_BriefFossilFuelSubsidies_EN_Web.pdf

225. United States Climate Action Partnership. (2009). *A blueprint for legislative action.*

226. United States Climate Action Partnership. (2007). *A call for action, consensus principles and recommendations.*

227. US Department of Labor. *Labor force statistics from the current population survey.* Available from: http://data.bls.gov/timeseries/LNS14000000

228. US Environmental Protection Agency (EPA). (2009). *Endangerment and cause or contribute findings for greenhouse gases under Section 202 (a) of the Clean Air Act. Federal Register, 74*(239), 66496–66546.

229. US Environmental Protection Agency. *U.S. Greenhouse gas emissions in 2013.* Available from: http://www3.epa.gov/climatechange/ghgemissions/gases.html

230. United States of America. *US intended nationally determined contribution.* Available from: http://www4.unfccc.int/submissions/INDC/Submission%20Pages/submissions.aspx

231. US Department of Commerce. (2009, June 25). *Gross domestic product, 1st Quarter 2009 (final).* [Press Release].

232. US Department of Commerce. (2009, February 27). *Gross domestic product: 4th Quarter 2008 (final).* [Press Release].

233. US Department of State. *Summary: North American 2015 HFC submission to the Montreal protocol.* Available from: http://www.state.gov/documents/organization/240964.pdf

234. US Department of State. (2014). *United States climate action report 2014, First Biennial report of the United States of America, Sixth National Communication of the United States of America, under the United Nations framework convention on climate change.* Washington, DC: US Department of State.

235. US Department of State. (2016). *United States climate action report 2016, Second Biennial report of the United States of America, under the United Nations framework convention on climate change.* Washington, DC: US Department of State.

236. US Draft Protocol Framework. (1997).

237. US Energy Information Administration. (2015). *Annual energy outlook 2015 with projections till 2040.* Washington, DC: Energy Information Administration. DOE/EIA-0383.

238. US Energy Information Administration. (1998). *Impacts of the Kyoto protocol on U.S. energy markets and economic activity.* Washington, DC: Energy Information Administration. SR/OIAF/98-03.

239. US Environmental Protection Agency. 40 CFR Chapter 1. *Endangerment and cause or contribute findings for greenhouse gases under Section 202(a) of the Clean Air Act.* Available from: http://www3.epa.gov/climatechange/Downloads/endangerment/Federal_Register-EPA-HQ-OAR-2009-0171-Dec.15-09

240. US Environmental Protection Agency. (2015). *Emission guidelines, compliance times, and standards of performance for municipal solid waste landfills*; proposed rules. Washington, DC. 40 CFR Part 60.

241. US Environmental Protection Agency. (2009). *EPA analysis of the American Clean Energy and Security Act of 2009.* Washington, DC: US Environmental Protection Agency.

242. US Energy Information Administration. (2015). *December 2015 Monthly energy review. Carbon dioxide emissions from energy consumption: Electric power sector.* US Department of Energy. DOE/EIA-0035(2015/12).

243. US Energy Information Administration. *Energy explained: How much coal is left?* Available from: https://www.eia.gov/energyexplained/index.cfm?page=coal_reserves

244. US Energy Information Administration. *Frequently asked questions: How old are U.S. nuclear power plants and when was the last one built.* Available from: https://www.eia.gov/tools/faqs/faq.cfm?id=228&t=21

245. US Environmental Protection Agency. *Acid rain program benefits exceed expectation.* Available from: http://www.epa.gov/capandtrade/documents/benefits.pdf. [No date of publication provided].

246. US Environmental Protection Agency. *Global greenhouse gas emissions.* Available from: http://www3.epa.gov/climatechange/science/indicators/ghg/global-ghg-emissions.html

247. US Environmental Protection Agency. *Overview of greenhouse gas emissions.* Available from: http://www3.epa.gov/climatechange/ghgemissions/gases.html

248. US Environmental Protection Agency. *Sources of greenhouse gas emissions.* Available from: http://www3.epa.gov/climatechange/ghgemissions/sources.html

249. US Environmental Protection Agency. (2015). *Inventory of U.S. greenhouse gas emissions and sinks: 1990–2013.* EPA 430-R-15-004.

250. US Environmental Protection Agency. (2011). *NOx budget trading program/NOx SIP call, 2003–2008.* Available from: http://www.epa.gov/airmarkets/progsregs/nox/sip.html. Last updated 2011.

251. US Environmental Protection Agency, & US Department of Transportation. (2010). *Light-duty vehicle greenhouse gas emissions standards and corporate average fuel economy standards*; final rule. Washington, DC. 40 CFR Parts 85, 86, and 600.

252. US Environmental Protection Agency, US Department of Transportation. (2012). *2017 and later model year light duty vehicle greenhouse gas emissions standards and corporate average fuel economy standards*; final rule. Washington, DC. 40 CFR Parts 85, 86, and 600.

253. US Environmental Protection Agency, US Department of Transportation. (2011). *Greenhouse gas efficiency standards and fuel efficiency standards for medium-and heavy-duty engines and vehicles*; final rule. Washington, DC. 40 CFR Parts 85, 86, 600, 1033, 1036, 1037, 1039, 1065, 1066, and 1068.

254. US EPA, US Department of Transportation. (2015). *Greenhouse gas emissions and fuel efficiency standards for medium and heavy-duty engines and vehicles—Phase 2*; proposed rule. 40 CFR Parts 9, 22, 85, 86, 600, 1033, 1036, 1037, 1039, 1042, 1043, 1065, 1066, and 1068. Washington, DC.

255. US Environmental Protection Agency. (2015). *Carbon pollution emission guidelines for existing stationary sources: Electric utility generating units,* final rule. Washington, DC. 40 CFR Part 60.

256. US Environmental Protection Agency. *EPA and NHTSA adopt first-ever program to reduce greenhouse gas emissions and improve fuel efficiency of medium- and heavy duty vehicles.* Available from: http://www3.epa.gov/otaq/climate/documents/420f11031.pdf

257. US Environmental Protection Agency. (2015). *Fast facts: US transportation sector GHG emissions 1990–2011.* Office of transportation and air quality. EPA-420-F-15-032.

258. US Environmental Protection Agency. (2015). Oil and natural gas sector: *Emissions standards for new and modified sources,* proposed rule. Washington, DC. 40 CFR Part 60.

259. US Environmental Protection Agency. *SNAP regulations.* Available from: http://www.epa.gov/snap/snap-regulations

260. US Environmental Protection Agency. (2015). *Proposed finding that greenhouse gas emissions from aircraft cause or contribute to air pollution that may reasonably be anticipated to endanger public health and welfare and advance notice of proposed rulemaking.* Washington, DC. 40 CFR Parts 87 and 1068.

261. US Environmental Protection Agency. *Regulation and standards: Light duty.* Available from: http://www3.epa.gov/otaq/climate/regs-light-duty.htm

262. US Environmental Protection Agency. *Sources of greenhouse gas emissions. Estimates from the inventory of US greenhouse gas emissions and sinks: 1990–2013.* Available from: http://www3.epa.gov/climatechange/ghgemissions/sources/electricity.html

263. US Environmental Protection Agency. (2015). *Standards of performance for greenhouse gas emissions from new, modified, and reconstructed stationary sources: Electric utility generating units;* final rule. Washington, DC. 40 CFR Parts 60, 70, 71, and 98.

264. Vice President Biden. *Progress report: The transformation to a clean energy economy.* Available from: https://www.whitehouse.gov/administration/vice-president-biden/reports/progress-report-transformation-clean-energy-economy

265. Victor, D. G. (2011). *Global warming gridlock: Creating more effective strategies for protecting the planet.* New York: Cambridge University Press.

266. Wara, M. (2008). Measuring the clean development mechanism's performance and potential. *UCLA Law Review, 56,* 1759–1803.

267. Weiss, D. (2010). Anatomy of a Senate climate bill death. This material [article] was created by the Center for American Progress.

268. Western Climate Initiative. *History.* Available from: http://www.westernclimateinitiative.org/history

269. The White House Office of the Press Secretary. *US-China joint announcement on climate change.* Available from: https://www.whitehouse.gov/the-press-office/2014/11/11/us-china-joint-announcement-climate-change

270. The White House Office of the Press Secretary. *U.S.-China joint presidential statement on climate change.* Available from: https://www.whitehouse.

gov/the-press-office/2015/09/25/us-china-joint-presidential-statement-climate-change

271. The White House Office of the Press Secretary. *Fact sheet: US and India Climate and Clean Energy Cooperation*. Available from: https://www.whitehouse.gov/the-press-office/2015/01/25/fact-sheet-us-and-india-climate-and-clean-energy-cooperation

272. The White House Office of the Press Secretary. (2015, June 30). *US-Brazil joint statement on climate change*. Available from: https://www.whitehouse.gov/the-press-office/2015/06/30/us-brazil-joint-statement-climate-change

273. Whitley, S. (2013). *Time to change the game: Fossil fuel subsidies and climate*. London: Overseas Development Institute.

274. Witte, G. (2015, November 21). Britain pulls the plug on renewable energy. *The Washington Post*.

275. The World Bank. *Prototype carbon fund*. Available from: http://www.worldbank.org/en/topic/climatechange/brief/world-bank-carbon-funds-facilities

276. The World Bank. *BioCarbon fund*. Available from: http://www.worldbank.org/en/topic/climatechange/brief/world-bank-carbon-funds-facilities

277. The World Bank. *Community development carbon fund*. Available from: http://www.worldbank.org/en/topic/climatechange/brief/world-bank-carbon-funds-facilities

278. The World Bank. *Statement September 12, 2013*. Available from: https://wbcarbonfinance.org/Router.cfm?Page=UCF

279. The World Bank. (2006, August 30). *Umbrella carbon facility completes allocation of first tranche*. [Press Release].

280. World Bank. (2014). *State and trends of carbon pricing 2014*. Washington, DC: World Bank. doi:10.1596/978-1-4648-0268-3

281. World Resources Institute. *CAIT climate data explorer: Paris contributions map*. Available from: http://cait.wri.org/indc/

282. Youngman, R., & Rosenzweig, R. (2008). Using offset systems to stimulate large-scale GHG reductions. In: *Greenhouse gas market 2008: Piecing together a comprehensive international agreement for a truly global carbon market*. Geneva: International Emissions Trading Association.

283. Zindler, E., Di Capua. M., et al. (2015). *Factbook: Sustainable energy in America*. New York: Bloomberg New Energy Finance, The Business Council for Sustainable Energy.

Index

© The Author(s) 2016

311

R.H. Rosenzweig, *Global Climate Change Policy and
Carbon Markets*, Energy, Climate and the Environment,
DOI 10.1057/978-1-137-56051-3

Printed in the United States
By Bookmasters